Role of Food Digestion and Digestive System in the Nutritional, Functional and Health Properties of Food Bioactives

Role of Food Digestion and Digestive System in the Nutritional, Functional and Health Properties of Food Bioactives

Guest Editor

Samuel Fernández-Tomé

MDPI

Basel • Beijing • Wuhan • Barcelona • Belgrade • Novi Sad • Cluj • Manchester

Guest Editor
Samuel Fernández-Tomé
Department of Nutrition and
Food Science
Complutense University of
Madrid (UCM)
Madrid
Spain

Editorial Office
MDPI AG
Grosspeteranlage 5
4052 Basel, Switzerland

This is a reprint of the Special Issue, published open access by the journal *Nutrients* (ISSN 2072-6643), freely accessible at: https://www.mdpi.com/journal/nutrients/special_issues/Food_Digestion_and_Digestive_System.

For citation purposes, cite each article independently as indicated on the article page online and as indicated below:

Lastname, A.A.; Lastname, B.B. Article Title. *Journal Name* **Year**, *Volume Number*, Page Range.

ISBN 978-3-7258-6035-7 (Hbk)
ISBN 978-3-7258-6036-4 (PDF)
https://doi.org/10.3390/books978-3-7258-6036-4

Contents

About the Editor

Samuel Fernández-Tomé

Samuel Fernández-Tomé has a degree in Veterinary with honors from Complutense University of Madrid (UCM, 2011) and MSc at the Department of Nutrition, Bromatology and Food Technology (UCM, 2012). PhD in Biology and Food Science focused on the gastrointestinal digestion of food proteins, dietary bioactive peptides, and their effects on digestive health, at the Institute of Food Science Research (CIAL, CSIC-UAM, 2012–2016) and granted a competitive PhD fellowship from the Spanish Ministry of Economy and Competitiveness. PhD internships (ICTAN-CSIC, URJC, IdiPAZ, and EEZ-CSIC from Spain; and University of Massachusetts from the USA, granted by the International Mobility FPI-Program). Postdoctoral stage at the Health Research Institute of Hospital University La Princesa in the Inflammatory Bowel Disease Unit, granted by the competitive Sara Borrell postdoctoral fellowship (Instituto de Salud Carlos III, 2017–2021). Participation in more than 20 competitive research projects (1 as PI), and contribution to more than 50 national or international conferences. Accredited as Assistant Professor and Associate Professor by The National Agency for Quality Assessment and Accreditation of Spain (ANECA). In November 2021, incorporated into the Department of Nutrition and Food Science (Faculty of Pharmacy, UCM). Teaching responsibilities at UCM in the Pharmacy, Food Science and Technology, and Human Nutrition and Dietetics degrees, as well as in the Dual degree, Pharmacy-Human Nutrition and Dietetics. Multidisciplinary research activity mainly focused on the role of food-derived bioactive compounds at the gastrointestinal level.

nutrients

Editorial

Role of Food Digestion and Digestive System in the Nutritional, Functional and Health Properties of Food Bioactives

Samuel Fernández-Tomé

Department of Nutrition and Food Science, Faculty of Pharmacy, Complutense University of Madrid (UCM), Plaza Ramón y Cajal s/n, 28040 Madrid, Spain; sfernandeztome@ucm.es

The impact of food components on the human digestive system is an important area of research in the fields of nutrition and food science. Food is composed of a mix of chemical components that will be processed by the digestive system to release, absorb, and metabolize the nutrients. The composition and structure of food play a significant role in its nutritional and functional properties, making it necessary to understand the modifications and mechanisms occurring during food digestion to better comprehend the potential health benefits of food bioactives. Furthermore, the human gastrointestinal tract is continually exposed to a varied range of food compounds in the gut lumen as they are processed following oral intake and then transported across the intestinal barrier. Food compounds may undergo several changes in the digestive system; hence, it is a real challenge for food scientists to understand their health implications within and beyond the gastrointestinal tract. Researchers have thus utilized some in vitro models in an attempt to simulate the gastrointestinal digestion process and study the interaction between foods and the digestive system, but the complexity of this interface require further investigation to advance our understanding of food digestion, bioaccessibility, and bioavailability, along with the influence of gut microbiota. The field of functional foods aims to uncover these complexities on the path towards enhancing our knowledge on the impact of food bioactives on human health and disease.

This Special Issue of the journal *Nutrients* includes eight outstanding papers describing examples of the new research trends on the interplay between the following key elements in nutrition and food science research: (i) the physiology of the digestive system and gastrointestinal digestion of foods; (ii) the analysis of the food matrix and composition during digestion; (iii) the nutritional, functional, and biological properties of food bioactives; and (iv) the health impact of the modulation of digestive functions through dietary substances.

The Special Issue begins with a group of research articles investigating different aspects related to the digestibility of food proteins (Contributions 1–4). The study of Benede et al. (Contribution 1) aimed at evaluating the susceptibility to digestion of transglutaminase-mediated crosslinked skim milk and its two main protein fractions, namely casein and whey protein, as well as their allergenic potential. After treatment with transglutaminase, the human IgE binding capacity of sera from milk-allergic patients to whey protein was increased but reduced in the case of skim milk and casein. Despite no effects being provoked in mast cell degranulation, in comparison to native proteins, the cross-linking of skim milk and casein induced a lower release of Th2 cytokines from the splenocytes of sensitized mice. Moreover, all transglutaminase-treated milk samples were more resistant to simulated digestion, and, consequently, the binding to human IgE of their digests was increased. This fact questioned the use of transglutaminase-crosslinking to obtain hypoallergenic products for cow's milk allergic patients and highlighted the necessity to evaluate the whole food matrix composed of diverse compounds with different resistance to processing and gastrointestinal digestion and, therefore, showing different interactions with gut mucosa and allergenicity mediators [1,2].

Nowadays, it is a well-known fact that the COST Action INFOGEST was established as an international network with the final aim of "Improving health properties of food by sharing our knowledge on the digestive process". It has been working on food digestion for several years and currently gathers more than 440 researchers from more than 40 countries worldwide. Hence, the network reached a consensus to harmonize the in vitro static systems that simulate the digestive processes by defining key parameters and conditions, hence allowing the publication of the standardized static in vitro digestion method suitable for food developed by Minekus et al. [3]. Noteworthy, the methodology achieved an extraordinary impact in the scientific community, currently reaching over 3300 citations since its origin from multiple research institutions worldwide, and it has been amended in an updated publication of the improved digestion protocol (INFOGEST 2.0) [4]. In this Special Issue, the study of Correa et al. (Contribution 2) applied this simulated digestion method to evaluate the release of essential amino acids and bioactive peptides from the legume *Erythrina edulis (chachafruto)* proteins during gastrointestinal digestion. Legume-derived food peptides have been shown in the literature to display mostly antihypertensive, antimicrobial, antidiabetic, or antiproliferative properties [5]. In this specific study on chachafruto proteins, low molecular weight peptides released after protein digestion were proved to exert antioxidant activity through radical scavenging capacities, whereas high molecular weight peptides prompted both immune activation in RAW 264.7 macrophages and cell protection against LPS challenge.

In the study of Jiménez-Muñoz et al. (Contribution 3), a semi-dynamic INFOGEST-based digestion system provided useful information about optimal gastrointestinal conditions for protein bioaccessibility, but also about the mechanistic understanding of how food matrix may influence the behavior of proteins during gastric and intestinal digestion. Hence, the study made use of confocal laser scanning microscopy, and protein breakdown and free amino acid analyses to demonstrate how the differences in the structure of four isocaloric matrices (suspension, gel, foam, and heat-set foam) of a commercial potato protein ingredient impact its digestibility and therefore its protein quality [6]. A greater structural complexity was found to lead to more delayed gastric emptying. In addition to a slower gastric emptying rate, the heat-treated samples revealed a higher degree of hydrolysis and lower trypsin inhibitory activity than the non-heat-treated samples. Regarding the distribution of the bioaccessible peptides formed, similar trends were found between samples at different gastric and intestinal times, apart from the heat-set foam matrix.

Following this line of research, the study of Jiménez-Barrios et al. (Contribution 4) characterized in detail the peptidome profile of duodenal digests from pigs as a recognized model of human digestion after oral administration of two dairy samples: micellar casein and a previously described casein hydrolysate [7]. When animals received micellar casein, a slower transit of nitrogen to the duodenum was observed, and the duodenal digests contained a wider range of peptide sizes as well as a higher number of sequences longer than five amino acids. Likewise, noticeable differences were revealed between the sequences identified in casein and hydrolysate digests, including the presence of different opioid peptides. Furthermore, the amino acid availability was quantified in plasma and evidenced a more rapid absorption in the hydrolysate group, with several branched-chain and other hydrophobic amino acids peaking at shorter times in comparison to the casein-fed animals.

Specifically focused on food proteins (Contributions 1–4), these investigations represent not only a key tool to validate in vitro digestion protocols but also allow for the obtention of valuable data that could be applied to physiological and metabolic research studies [8,9]. The fate of food proteins during gastrointestinal digestion and the biological effects of their corresponding released peptides are thus highlighted as an area of intense research in the field of functional foods [10,11]. Hence, food-derived peptides with biological properties at the gastrointestinal level, such as antioxidant and antimicrobial [12], modulation of inflammation and immune response [13], chemoprevention of malignant diseases [14], opioid activity [15], or anti-diabetic effects [16], among others, have been reported in various investigations.

Food researchers have also utilized gastrointestinal digestion protocols to evaluate the bioaccessibility and bioavailability of nutrients and bioactive compounds [17]. In this respect, the study of Fărcaș et al. (Contribution 5) used brewers' spent grain, the main by-product obtained from the industrial brewing process, as a sustainable source of minerals and B group vitamins and evaluated its bioaccessibility. The study found that micro-minerals such as Fe, Cu, Zn, Cr, Mn, Ni, Ba, and Sr and macro-minerals such as Ca, Mg, Na, K, and B group vitamins (B1, B3, B6, and B12) presented different bioaccessibility values following the INFOGEST digestion methodology. Indeed, these authors suggested that the diverse sample types (including different percentages of malted cereals), their chemical composition (protein, fiber, bound polypeptide, and polysaccharide contents), along with the gastrointestinal digestion conditions, were the main factors that finally affected the bioaccessibility values of the assessed minerals and B vitamins.

Regarding bioactive compounds, proanthocyanidins are a type of polyphenol with previously described preventive properties against obesity alterations [18]. In the study of Yu et al. (Contribution 6), persimmon proanthocyanidins significantly reduced the body weight of high-fat-diet-fed C57BL/6J mice, with different improvements in insulin resistance, lipid accumulation, as well as gut microbiota composition and diversity depending on their degrees of polymerization. Furthermore, in comparison with the oligomeric proanthocyanidins, polymeric proanthocyanidins exhibited strong inhibition over the activity of the digestive enzymes α-amylase and pancreatic lipase. In this line of research, previous studies had shown encouraging results for the inhibition of lipid digestion enzymes using bioactive compounds derived from black tea brew and grape seed powder [19,20]. Hence, the study of Tormási and Abrankó (Contribution 7) examined the lipolysis inhibitory effect of both ingredients when co-digested with two selected real foods with different lipid profiles as substrate tests (cream and baked beef). To that purpose, this study also applied the INFOGEST digestion protocol. The authors found that the characteristics of the dietary fat source can influence the lipolysis process, as both samples similarly decreased the lipolysis of the cream sample, but they were not able to influence the digestibility of beef fat, showing a simpler fatty acid profile in comparison with the diverse fatty acid composition of milk fat. Moreover, it was suggested that the lipolysis inhibitory effect of black tea brew and grape seed powder primarily affected the lipolysis of short- and medium-chain fatty acids because of the different preference of pancreatic lipase, in opposition to gastric lipase, to digest this type of triacylglycerol, which is further enhanced as a result of co-digestion with the evaluated ingredients.

Lastly, the Special Issue concludes with a review by Paterson et al. (Contribution 8) that summarizes the novel trends in microalgae research. This comprehensive review provides an updated discussion on the nutritional value (protein and amino acids, lipids and fatty acids, carbohydrates, minerals, vitamins, carotenoids, and phenolic compounds), biological effects (antioxidant, antimicrobial, and anticarcinogenic), as well as the digestibility of two microalgae genera, *Tetraselmis* and *Nannochloropsis*, as the basis to support their alternative exploitation as ingredients with health benefits and, consequently, with potential for the development of novel functional foods.

Acknowledgments: The invited authors are acknowledged for their interesting and insightful contributions.

Conflicts of Interest: The authors declare no conflicts of interest.

List of Contributions:

1. Benedé, S.; Martínez-Blanco, M.; López-Fandiño, R.; Molina, E. IgE-Binding and Immunostimulating Properties of Enzymatic Crosslinked Milk Proteins as Influenced by Food Matrix and Digestibility. *Nutrients* **2022**, *14*, 4584. https://doi.org/10.3390/nu14214584.
2. Correa, J.L.; Zapata, J.E.; Hernández-Ledesma, B. Release of Bioactive Peptides from *Erythrina edulis* (*Chachafruto*) Proteins under Simulated Gastrointestinal Digestion. *Nutrients* **2022**, *14*, 5256. https://doi.org/10.3390/nu14245256.

3. Jiménez-Munoz, L.; Tsochatzis, E.D.; Corredig, M. Impact of the Structural Modifications of Potato Protein in the Digestibility Process under Semi-Dynamic Simulated Human Gastrointestinal In Vitro System. *Nutrients* **2022**, *14*, 2505. https://doi.org/10.3390/nu14122505.
4. Jiménez-Barrios, P.; Sánchez-Rivera, L.; Martínez-Maqueda, D.; Le Gouar, Y.; Dupont, D.; Miralles, B.; Recio, I. Peptidomic Characterization and Amino Acid Availability after Intake of Casein vs. a Casein Hydrolysate in a Pig Model. *Nutrients* **2023**, *15*, 1065. https://doi.org/10.3390/nu15051065
5. Fărcaș, A.C.; Socaci, S.A.; Chiș, M.S.; Martínez-Monzó, J.; García-Segovia, P.; Becze, A.; Török, A.I.; Cadar, O.; Coldea, T.E.; Igual, M. *In Vitro* Digestibility of Minerals and B Group Vitamins from Different Brewers' Spent Grains. *Nutrients* **2022**, *14*, 3512. https://doi.org/10.3390/nu14173512.
6. Yu, Y.; Chen, P.; Li, X.; Shen, S.; Li, K. Persimmon Proanthocyanidins with Different Degrees of Polymerization Possess Distinct Activities in Models of High Fat Diet Induced Obesity. *Nutrients* **2022**, *14*, 3718. https://doi.org/10.3390/nu14183718.
7. Tormási, J.; Abrankó, L. Impact of Grape Seed Powder and Black Tea Brew on Lipid Digestion—An In Vitro Co-Digestion Study with Real Foods. *Nutrients* **2023**, *15*, 2395. https://doi.org/10.3390/nu15102395.
8. Paterson, S.; Gómez-Cortés, P.; de la Fuente, M.A.; Hernández-Ledesma, B. Bioactivity and Digestibility of Microalgae *Tetraselmis* sp. and *Nannochloropsis* sp. as Basis of Their Potential as Novel Functional Foods. *Nutrients* **2023**, *15*, 477. https://doi.org/10.3390/nu15020477.

References

1. Verhoeckx, K.C.M.; Vissers, Y.M.; Baumert, J.L.; Faludi, R.; Feys, M.; Flanagan, S.; Herouet-Guicheney, C.; Holzhauser, T.; Shimojo, R.; van der Bolt, N.; et al. Food processing and allergenicity. *Food Chem. Toxicol.* **2015**, *80*, 223–240. [CrossRef] [PubMed]
2. Verhoeckx, K.; Bøgh, K.L.; Dupont, D.; Egger, L.; Gadermaier, G.; Larré, C.; Mackie, A.; Menard, O.; Adel-Patient, K.; Picariello, G.; et al. The relevance of a digestibility evaluation in the allergenicity risk assessment of novel proteins. Opinion of a joint initiative of COST action ImpARAS and COST action INFOGEST. *Food Chem. Toxicol.* **2019**, *129*, 405–423. [CrossRef] [PubMed]
3. Minekus, M.; Alminger, M.; Alvito, P.; Ballance, S.; Bohn, T.; Bourlieu, C.; Carrière, F.; Boutrou, R.; Corredig, M.; Dupont, D.; et al. A standardised static in vitro digestion method suitable for food-an international consensus. *Food Funct.* **2014**, *5*, 1113–1124. [CrossRef] [PubMed]
4. Brodkorb, A.; Egger, L.; Alminger, M.; Alvito, P.; Assunção, R.; Ballance, S.; Bohn, T.; Bourlieu-Lacanal, C.; Boutrou, R.; Carrière, F.; et al. INFOGEST static in vitro simulation of gastrointestinal food digestion. *Nat. Protoc.* **2019**, *14*, 991–1014. [CrossRef] [PubMed]
5. Luna-Vital, D.; de Mejía, E.G. Peptides from Legumes with Antigastrointestinal Cancer Potential: Current Evidence for Their Molecular Mechanisms. *Curr. Opin. Food Sci.* **2018**, *20*, 13–18. [CrossRef]
6. Herreman, L.; Nommensen, P.; Pennings, B.; Laus, M.C. Comprehensive overview of the quality of plant-and animal sourced proteins based on the digestible indispensable amino acid score. *Food Sci. Nutr.* **2020**, *8*, 5379–5391. [CrossRef] [PubMed]
7. Contreras, M.d.M.; Sevilla, M.A.; Monroy-Ruiz, J.; Amigo, L.; Gómez-Sala, B.; Molina, E.; Ramos, M.; Recio, I. Food-Grade Production of an Antihypertensive Casein Hydrolysate and Resistance of Active Peptides to Drying and Storage. *Int. Dairy J.* **2011**, *21*, 470–476. [CrossRef]
8. Boutrou, R.; Gaudichon, C.; Dupont, D.; Jardin, J.; Airinei, G.; Marsset-Baglieri, A.; Benamouzig, R.; Tomé, D.; Leonil, J. Sequential Release of Milk Protein Derived Bioactive Peptides in the Jejunum in Healthy Humans. *Am. J. Clin. Nutr.* **2013**, *97*, 1314–1323. [CrossRef] [PubMed]
9. Egger, L.; Schlegel, P.; Baumann, C.; Stoffers, H.; Guggisberg, D.; Brügger, C.; Dürr, D.; Stoll, P.; Vergères, G.; Portmann, R. Physiological Comparability of the Harmonized INFOGEST In Vitro Digestion Method to In Vivo Pig Digestion. *Food Res. Int.* **2017**, *102*, 567–574. [CrossRef] [PubMed]
10. Fernández-Tomé, S.; Hernández-Ledesma, B. Gastrointestinal digestion of food proteins under the effects of released bioactive peptides on digestive health. *Mol. Nutr. Food Res.* **2020**, *64*, 2000401. [CrossRef] [PubMed]
11. Duarte Villas Mishima, M.; Stampini Duarte Martino, H.; Silva Meneguelli, T.; Tako, E. Effect of food derived bioactive peptides on gut health and inflammatory mediators in vivo: A systematic review. *Crit. Rev. Food Sci. Nutr.* **2023**, 1–11. [CrossRef]
12. López-García, G.; Dublan-García, O.; Arizmendi-Cotero, D.; Gómez Oliván, L.M. Antioxidant and antimicrobial peptides derived from food proteins. *Molecules* **2022**, *27*, 1343. [CrossRef] [PubMed]
13. Fernández-Tomé, S.; Indiano-Romacho, P.; Mora-Gutiérrez, I.; Pérez-Rodríguez, L.; Ortega Moreno, L.; Marin, A.C.; Baldán-Martín, M.; Moreno-Monteagudo, J.A.; Santander, C.; Chaparro, M.; et al. Lunasin peptide is a modulator of the immune response in the human gastrointestinal tract. *Mol. Nutr. Food Res.* **2021**, *65*, 2001034. [CrossRef] [PubMed]
14. Díaz-Gómez, J.L.; Castorena-Torres, F.; Preciado-Ortiz, R.E.; García-Lara, S. Anti-cancer activity of maize bioactive peptides. *Front. Chem.* **2017**, *5*, 44. [CrossRef] [PubMed]

15. Liu, Z.; Udenigwe, C.C. Role of food-derived opioid peptides in the central nervous and gastrointestinal systems. *J. Food Biochem.* **2019**, *43*, e12629. [CrossRef] [PubMed]
16. Elam, E.; Feng, J.; Lv, Y.M.; Ni, Z.J.; Sun, P.; Thakur, K.; Zhang, J.G.; Ma, Y.L.; Wei, Z.J. Recent advances on bioactive food derived anti-diabetic hydrolysates and peptides from natural resources. *J. Funct. Foods* **2021**, *86*, 104674. [CrossRef]
17. Grundy, M.M.; Moughan, P.J.; Wilde, P.J. Bioaccessibility and associated concepts: Need for a consensus. *Trends Food Sci. Technol.* **2024**, *145*, 104373. [CrossRef]
18. Zhu, W.; Oteiza, P.I. Proanthocyanidins at the gastrointestinal tract: Mechanisms involved in their capacity to mitigate obesity-associated metabolic disorders. *Crit. Rev. Food Sci. Nutr.* **2024**, *64*, 220–240. [CrossRef] [PubMed]
19. Dwibedi, V.; Jain, S.; Singhal, D.; Mittal, A.; Rath, S.K.; Saxena, S. Inhibitory Activities of Grape Bioactive Compounds against Enzymes Linked with Human Diseases. *Appl. Microbiol. Biotechnol.* **2022**, *106*, 1399–1417. [CrossRef] [PubMed]
20. Chen, T.-Y.; Wang, M.M.C.; Hsieh, S.-K.; Hsieh, M.-H.; Chen, W.-Y.; Tzen, J.T.C. Pancreatic Lipase Inhibition of Strictinin Isolated from Pu'er Tea (*Cammelia sinensis*) and Its Anti-Obesity Effects in C57BL6 Mice. *J. Funct. Foods* **2018**, *48*, 1–8. [CrossRef]

nutrients

MDPI

Review

Bioactivity and Digestibility of Microalgae *Tetraselmis* sp. and *Nannochloropsis* sp. as Basis of Their Potential as Novel Functional Foods

Samuel Paterson, Pilar Gómez-Cortés *, Miguel Angel de la Fuente and Blanca Hernández-Ledesma *

Department of Bioactivity and Food Analysis, Institute of Food Science Research (CIAL, CSIC-UAM, CEI UAM+CSIC), Nicolás Cabrera 9, 28049 Madrid, Spain
* Correspondence: p.g.cortes@csic.es (P.G.-C.); b.hernandez@csic.es (B.H.-L.)

Abstract: It is estimated that by 2050, the world's population will exceed 10 billion people, which will lead to a deterioration in global food security. To avoid aggravating this problem, FAO and WHO have recommended dietary changes to reduce the intake of animal calories and increase the consumption of sustainable, nutrient-rich, and calorie-efficient products. Moreover, due to the worldwide rising incidence of non-communicable diseases and the demonstrated impact of diet on the risk of these disorders, the current established food pattern is focused on the consumption of foods that have functionality for health. Among promising sources of functional foods, microalgae are gaining worldwide attention because of their richness in high-value compounds with potential health benefits. However, despite the great opportunities to exploit microalgae in functional food industry, their use remains limited by challenges related to species diversity and variations in cultivation factors, changes in functional composition during extraction procedures, and limited evidence on the safety and bioavailability of microalgae bioactives. The aim of this review is to provide an updated and comprehensive discussion on the nutritional value, biological effects, and digestibility of two microalgae genera, *Tetraselmis* and *Nannochloropsis*, as basis of their potential as ingredients for the development of functional foods.

Keywords: microalgae; nutritional value; bioactivity; digestibility; novel functional foods

1. Introduction

Microalgae are single-celled, prokaryotic, and eukaryotic primary photosynthetic microorganisms with diverse taxonomical and phylogenetical characteristics. Their distinguished ability to produce biomass from CO_2, solar energy and elemental nutrients combined with their adaptation capacity to extreme surroundings have led to their presence in a myriad of environmental media throughout evolution [1,2]. Microalgae are classified on the basis of several features, such as the nature of the photosynthetic storage products, the organization of photosynthetic membranes, and the pigmentation, among others. The term microalgae includes different groups of eukaryotic organisms such as chlorophylls (green algae), bacillaryiophytes (diatoms), dinophytes (dinoflagellates), euglenophytes, prymne-siophytes (cocolitophorides) and prokaryotes, such ascaryophytes [2].

Tetraselmis sp. are green unicellular flagellates that belong to the family of *Chlorodendraceae* within the class of *Chlorodendrophyceae*, with almost spherical, slightly flattened cells with an invagination at the anterior end from which four equal flagella arise in two opposite pairs. Nowadays, numerous marine and freshwater species of this microalgae are known and some even occur in plankton [3]. *Tetraselmis* sp. have been used as a laboratory model strain to study the effect of diverse conditions such as saline or high nitrogen environments on their growth, survival and adaptation [4]. Moreover, these species are frequently used in aquaculture due to their high nutritional value in terms of proteins and fatty acids, and their content in other biological compounds [5]. Different strains of the *Tetraselmis marina* species have also been considered an innovative, cost-effective, efficient and sustainable method for the removal of toxic substances such as heavy metals from wastewater and aquatic ecosystems [6]. Recently, *Tetraselmis chuii* has been approved by the European Food Safety Authority as a novel food [7]. Moreover, the Scientific Committee of the Spanish Agency for Food Safety and Nutrition has declared that there is no indication that consumption of the *T. chuii* microalgae as a condiment can produce adverse effects on health [8].

Nannochloropsis sp. are unicellular, planktonic, with either 2–4 μm diameter subspherical or 3–4 × 1.5 μm cylindrical cells [9]. They belong to the family of *Monodopsidaceae* within the class of *Eustigmatophyceae*. *Nannochloropsis* sp. grow mainly in marine environments, and they normally contain a yellow-green chloroplast with the main pigments being chlorophyll A and xanthophylls [10]. Currently, seven *Nannochloropsis* sp. are recognized, six detected in marine environments (*N. salina, N. australis, N. granulata, N. oceanica, N. oculata,* and *N. gaditana*), and *N. limnetica,* which can be located in both fresh and brackish water [9]. These species of *Nannochloropsis* have become very popular for their use in animal feeding, aquaculture, and water waste treatment. In addition, their lipid production for biofuels and the production of long-chain polyunsaturated fatty acids (PUFAs), especially eicosapentaenoic acid (EPA, C20:5 n-3), are some of their main areas of interest to researchers [11,12]. Nevertheless, other different applications such as the development of natural cosmetics and cosmeceuticals are being currently explored [13].

Overall, *Tetraselmis* sp. and *Nannochloropsis* sp. microalgae provide environmental benefits, have a rich composition in nutrients and industrial applications as remarkable properties. However, the full applications and potential biotechnological roles of these microalgae in the food sector will directly depend on the bioactivity, digestibility, and bioavailability of their nutritional and bioactive compounds. Information about these genera is very abundant in the literature but it is rather scattered. This review aims to compile and update the composition, nutritional value, biological activity, and digestibility of *Tetraselmis* sp. and *Nannochloropsis* sp. microalgae biomass and derived compounds (Figure 1).

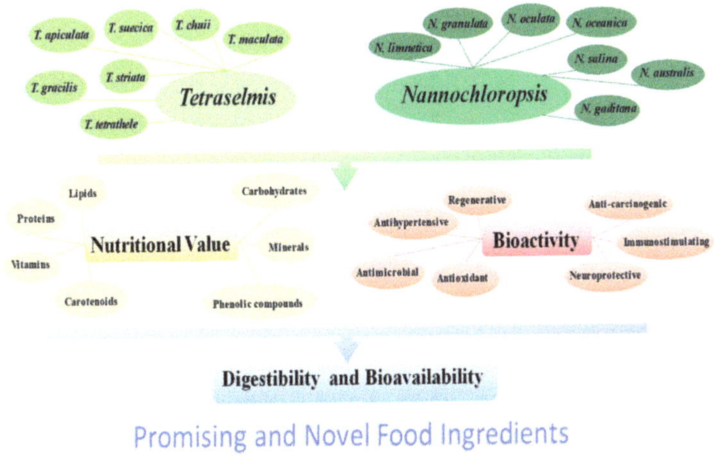

Figure 1. Overview of the nutritional value and bioactivities demonstrated for microalgae *Tetraselmis* sp. and *Nannochloropsis* sp. species.

2. Nutritional Value of *Tetraselmis* and *Nannochloropsis* sp. Microalgae

Microalgae are considered rich sources of macro- and micronutrients such as lipids, proteins, carbohydrates, minerals, and vitamins, among others. The biochemical composition of *Nannochloropsis* and *Tetraselmis* biomass show quantitative variations depending on the microalga strain as well as their growing environmental conditions. It is also remarkable that most of the existing composition data in literature come from microalgae cultivated on a laboratory scale, frequently different from industrial cultivation, thus affecting the nutrient profile.

2.1. Proteins and Amino Acids

Overall, *Nannochloropis* and *Tetraselmis* genera can be considered as a good source of proteins with desirable amino acid profile. Proteins in *Tetraselmis* range from 13% to 48% in dry weight (DW) biomass basis, although most species have protein levels of approximately 30%, mainly depending on the growing environmental factors [14–16]. The introduction of *Nannochloropsis* as a novel food is also justified by the high protein of appreciable nutritional value [17]. Protein content and its bioavailability are also affected by numerous biotic and abiotic factors and genomic variation among species. Values reported in the bibliography ranged from 19 to 24% for *N. oculata* [18] to 44.9% for *N. gaditana* [19]. Levasseur et al. [15] also compiled intermediates percentages for this genus.

The amino acid profile is important to assess the nutritional quality of foods. The quality of proteins can be determined specifically by the presence of essential amino acids (EAA). The amino acid composition of *Nannochloropsis* and *Tetraselmis* is listed in Table 1 together to reference values for EAA recommended by the Food and Agriculture Organization (FAO). Comparison of EAA profile of microalgae proteins with FAO reference pattern indicated that, overall, the EAA content of these microalgae is close to reference values that are indicative of good protein quality. However, methionine content is lower than stated by FAO. Microalgae also have a rich and varied composition of non-essential amino acids (Table 1), among which the most abundant are aspartate and glutamate (*Nannochloropsis*) and arginine (*Tetraselmis*). Other studies in *T. chuii* [20,21] confirm the quantitative importance of aspartic and glutamic acids in their proteins.

Table 1. Amino acid composition (expressed as g/16 g of nitrogen) of *Nannochloropsis* sp. and *Tetraselmis* sp. microalgae and comparison with recommended values by the Food and Agriculture Organization (FAO).

Amino Acid	Content (g/16 g Nitrogen)					
	FAO	*Nannochloropsis* [1]	*N. granulata* [2]	*N. oculata* [2]	*T. chuii* [2]	*T. suecica* [2]
Essential						
Isoleucine	4.0	3.6–4.2	5.6	4.8	3.5	3.5
Leucine	7.0	7.9–8.4	11.0	7.8	7.5	8.0
Valine	5.0	4.8–5.4	7.1	6.5	5.8	5.7
Lysine	5.5	5.2–6.0	8.5	6.1	5.7	6.0
Phenylalanine	6.0	4.4–4.8	6.2	6.2	5.4	5.9
Methionine	3.5	1.7–2.1	3.5	1.6	1.9	2.3
Tryptophan	1.0	1.5–1.8	2.8	1.6	3.7	3.8
Threonine	4.0	4.5–4.9	5.4	5.5	4.2	4.1
Non-essential						
Tyrosine		2.8–3.2	4.2	4.2	3.7	3.8
Cysteine		0.6–0.8	1.6	0.4	0.6	0.7
Alanine		6.2–6.9	7.1	7.4	6.8	6.9
Arginine		5.1–5.4	7.4	7.3	13.5	13.2
Aspartic acid		8.2–8.6	11.4	7.6	9.4	8.9
Glutamic acid		9.9–12.0	14.1	10.1	12.4	11.2
Glycine		5.1–5.4	7.5	5.5	5.9	5.9
Histidine		2.0–2.1	2.3	2.1	1.8	1.8
Proline		4.3–7.8	11.2	9.3	5.1	4.7
Serine		4.2–4.3	5.6	5.0	4.3	4.6
TAA		82.0–94.1	122.5	99.0	101.2	101.0
HAA		37.8–45.4	60.3	49.8	44.0	45.3
AAA		8.7–9.8	13.2	12.0	12.8	13.5

[1] Values from Wild et al. [22]; [2] values from Kumar et al. [14]. AAA: aromatic amino acids (phenylalanine, tryptophan, and tyrosine); HAA: hydrophobic amino acids (alanine, valine, isoleucine, leucine, tyrosine, phenylalanine, tryptophan, methionine, proline, and cysteine); TAA: total amino acids.

2.2. Lipids and Fatty Acids

The biomass of *Nannochloropsis* sp. is widely used as a source of healthy and essential lipids due to its high content in these molecules. *Nannochloropsis* has a lipid content (37 to 60% of DW) higher than that of other microalgae [23]. *Nannochloropsis* species are known to be rich in n-3 PUFAs, mainly EPA (C20:5 n-3), with contents ranging from 1.1% to 11% of its DW biomass [17]. Other dominant fatty acids are palmitic (16:0), palmitoleic (*cis*-9 16:1) and, with less abundance, oleic (*cis*-9 18:1) acids (Table 2). In contrast, the presence of lipids in species of *Tetraselmis* genus is clearly lower, between 5–10% DW [16,21,24]. The fatty acid profile is mainly composed of palmitic, oleic, linoleic (*cis*-9 *cis*-12 C18:2, n-6), and α-linolenic (*cis*-9 *cis*-12 *cis*-15 C18:3 n-3) acids. Other omega-3 PUFAs as EPA or C22:6 n-3 are less abundant (Table 2).

Table 2. Composition of fatty acids (expressed as mg/g of dry weight, DW) of *Nannochloropis* sp. and *Tetraselmis* sp. microalgae biomass.

Fatty Acid	Fatty Acids (mg/g DW)									
	N. oculata [1]	*N. limnetica* [2]	*N. gaditana* [3]	*N. granulata* [4]	*N. oceanica* [5]	*N. oceanica* [6]	*T.* sp. [5]	*T. suecica* [6]	*T. chuii* [4]	*T. chuii* [7]
SFA										
12:0	—	—	0.4	1.4	1.2	3.5	—	—	—	0.02
14:0	2.7	6.3	4.1	11.3	16.9	9.3	0.5	—	—	0.2
16:0	14.6	36.1	17.9	27.7	17.2	67.3	6.3	13.3	13.9	14.9
18:0		0.7	0.3	—	1.8	3.5	1.2	—	—	0.4
MUFA										
16:1	21.8	37.1	31.5	41.5	18.2	58.2	1.3	0.4	—	0.4
cis-9 18:1	6.3	30.1	1.9	11	4.1	30.6	10.7	19.5	6	18.4
20:1	—	—	—	—	0.5	—	0.9	—	1.2	1.7
PUFA										
16:2 n-4	—	—	—	—	—	—	—	—	1.8	—
18:2 n-6	2.4	3.8	1.4	7.7	9.7	2.8	2.5	11.9	6.2	4.8
18:3 n-6	—	0.2	0.8	—	—	0.7	—	—	—	0.5
20:2 n-6	—	—	—	—	0.5	0.4	—	—	—	0.04
20:4 n-6	5.8	5.3	3.2	—	3.7	4.8	0.6	—	—	0.4
18:3 n-3	—	0.4	0.1	—	0.5	0.2	6.4	17.6	10.8	9.7
18:4 n-3	—	—	—	—	—	—	4.1	—	—	—
20:5 n-3	23.3	28.1	54.9	31.7	23.4	31.5	4.8	—	3.8	4.1
22:6 n-3	—	0.1	0.20	—	—	—	0.2	—	—	—

[1] Volkman et al. [25]; [2] Krienitz and Wirtz [26]; [3] Verspreet et al. [27]; [4] Banskota et al. [28]; [5] Patil et al. [29]; [6] Niccolai et al. [30]; [7] Sandgruber et al. [21]. MUFA: monounsaturated fatty acids; PUFA: polyunsaturated fatty acids; SFA: saturated fatty acids.

The lipids of microalgae can be modified under experimental conditions. Illumination, nitrogenous (N):phosphorus (P) ratio and salinity are essential factors to influence the content and profile of lipid classes and fatty acids [23,31]. Furthermore, this behavior varies differently across species when exposed to the same culturing conditions. Cultures generally tend to accumulate lipids in slow cell proliferation phases, when most of the energy is available in reserve molecules, whereas EPA and other PUFAs increase with the cell density because of their role as cytoplasmic membrane components [32]. Van Wagenen et al. [33] postulated that the amount of PUFAs and EPA are inversely proportional to temperature, as physiological adaptations aimed to maintain the cell membrane fluidity.

Microalgae synthesize fatty acids mainly for esterification into glycerol-based membrane lipids but n-3 PUFAs can also be esterified to polar lipids such as phospholipids and glycolipids, which are more bioavailable [34]. These polar lipids can be abundant in the biomass of microalgae as *Nannochloropsis* [35,36] and *Tetraselmis* [37]. The *Nannochloropsis* sp. polar lipids have well-known bioactivities, including anti-inflammatory action [38]. Kagan et al. [39] demonstrated that polar lipid rich oil from *Nannochloropsis oculata* was an effective source of EPA in humans while Rao et al. [40] observed that supplementation with an EPA polar lipid extract of *Nannochloropsis* sp. increased this omega-3 PUFA as well as reduced the cholesterol level in healthy individuals.

2.3. Carbohydrates

Carbohydrates are nutrients also present in microalgae, both as intracellular storage molecules and cell wall structural elements. Zanella and Vianello [17] reported a range from 6.4 to 36.2% as fraction DW biomass in a variety of species of *Nannocholoropsis*. Carbohydrate's content can be manipulated through growth conditions. For instance, *T. suecica* was able to accumulate up to 37% under N-starved conditions, but when it was grown in nutrient replete medium, the carbohydrates content ranged from 6 to 10% [30]. With controlled growth conditions, *Tetraselmis* sp. carbohydrate contents were approximately 10% but under stressful conditions their values could exceed 50% [15].

Polysaccharide is the major form of carbohydrate in all species of microalgae. On average, in *Tetraselmis* and *Nannochloropsis*, approximately 90% of total carbohydrates are polysaccharides [41]. In *Nannochloropsis* sp., cellulose is the main component of its tough and highly recalcitrant cell wall [42]. These cell wall polysaccharides were first exploited for their rheological properties but could also be applied in the food and pharmaceuticals industries [15]. Species of *Nannochloropsis* genus may also be a source of β-glucans, glucose polymers linked by β-1,3, β-1,4 and branches of β-1,6 bonds [43]. β-glucans attract attention because of their bioactive properties and could grant a high potential to the culture of *Nannochloropsis* with commercial purposes [44].

The main intracellular polysaccharide described in *Tetraselmis* genus is starch [16], a storage macromolecule common in the vegetable word and, from a nutritional point of view, an important ingredient as source of energy. In addition, the cell wall polysaccharides of this genus have been reported to contain 3-deoxy-d-manno-oct-2-ulosonic acid, a molecule with pharmacological properties and a high economic value. This monosaccharide has been identified in *Tetraselmis striata*, *T. tetrathele* and *T. suecica* [45,46].

The sugar composition of the polysaccharides from microalgae differs between species and the variations could contribute to differences in the nutritional value, since animals digest polysaccharides of different composition at different rates. In *N. oculate*, glucose, the preponderant sugar content (68.2%) is followed by that of fucose, galactose, mannose, rhamnose, ribose and xylose in percentages below 10% [41]. A similar sugar pattern was reported by Templeton et al. [47] in *Nannochloropsis* sp. In *T. chuii* and *T. suecica*, glucose is also the principal sugar (>74%) while galactose, ribose, mannose, rhamnose, and arabinose were detected in lower percentages [41]. Pereira et al. [16] showed that *Tetraselmis* sp. CTP4 biomass is composed mainly of glucose too, including galactose, mannose arabinose and xylose contained in lower proportions.

2.4. Mineral Elements

Mineral data of *Nannochloropsis* and *Tetraselmis* genera are provided in Table 3. Major elements reported in these studies were Ca, Mg, K, Na, P and S. Minor and trace element compositions included Cu, Fe, Mn, Zn, and I. Furthermore, very low amounts of Se, Ni and Mo were also reported by some authors [21,22,48]. In *Nannochloropsis*, the richest elements detected were Na [22,42], K [48] and Ca [32], whereas Na [42], K [16], Ca [48] and Mg [21] were the richest detected in *Tetraselmis*. Iron was clearly the most abundant trace mineral in all these microalgae, with a concentration ranging from 30 to 177 mg/100 g of DW biomass.

There were very large differences in mineral elements concentrations between the species obtained from different studies (Table 3). Furthermore, the concentrations of minerals showed big differences, even within the same species. For instance, in *T. chuii*, the amount of Ca reported by Sandgruber et al. [21] was 15 times lower than that of *T. chuii* powder according to Tibbetts et al. [48]. This variability was greater for most of the trace elements contents and might result from different cultivation conditions, particularly the composition of the growing medium. Moreover, the concentration of the elements depends on the harvest of the microalgae in different growing stations in which elements were accumulated [21].

Table 3. Minerals (major and trace elements, expressed as mg/100 g of dry weight, DW) composition of *Nannochloropsis* sp. and *Tetraselmis* sp. microalgae.

Mineral	Content (mg/100 g DW)							
	N. sp. [1]	*N.* sp. [2]	*N. granulata* [3]	*N. gaditana* [4]	*T. chuii* [3]	*T. chuii* [4]	*T.* CTP4 [5]	*T. suecica* [6]
Major elements								
Calcium	972	590	90	631	2990	185	1190	1791
Magnesium	316	530	260	370	430	305	2080	601
Phosphorus		1300	730	887	1460		710	964
Potassium	533	1500	1500	1261	1860		4200	1349
Sodium	659	2500	1030	2022	890		1180	3223
Sulphur		610	580		1380			
Trace elements								
Copper	35	1.5	1. 8	2.9	10.2	0.5	1.1	1.1
Iron	136	62.7	139.5	48.9	177.4	51.8	32.3	30.9
Iodine		0.2				0.3	0.1	
Manganese	3.4	6.9	15.1	9.3	19.1	0.1		3.5
Selenium		0.01	0.1		0.1			
Zinc	103	5.5	3.2	10.1	6.4	2.1	2.9	6.6
Nickel						0.01		
Molybdenum						0.05		

[1] Rebolloso-Fuentes et al. [32]; [2] Wild et al. [22]; [3] Tibbetts et al. [48]; [4] Sandgruber et al. [21]; [5] Pereira et al. [16]; [6] Di Lena et al. [42].

2.5. Vitamins

Microalgae can synthesize and accumulate a wide range of vitamins and could be a potential source of these bioactive compounds compared to some well-known and traditional vegetables [15,49,50]. In Table 4, integrated state-of-the-art data on the vitamins A, C, D, E, K and B complex corresponding to *Nannochloropsis* and *Tetraselmis* genera can be found.

T. suecica presents the highest content of vitamin A and it seems to be rich in vitamins B1, B2, B3 and D2 [51,52]. Moreover, tocopherols values are especially high in this specie with concentrations up to 1.980 mg/100 g DW [53]. In *Nannochloropsis* genus, because of limited data, levels are still difficult to assess. Vitamin A was not detected or it was recorded in very low amount [54]. In contrast, *N. gaditana* powders have been shown to be a good source of vitamin B9 [55]. Its content (2.080 µg/100 g DW) represents approximately a quarter of the recommended daily intake (400 µg) for a consumption of 5 g of microalgal powder [15].

Table 4. Vitamin content in *Nannochloropsis* sp. and *Tetraselmis* sp. microalgae biomass.

Vitamin	Content (mg/100 g)											
	T. suecica [1]	*T. suecica* [2]	*T. suecica* [3]	*N.* sp. CS-246 [3]	*T.* sp. CS362 [4]	*T.* sp. CTP4 [5]	*N.* sp. [4]	*N. oculata* [6]	*N. limnetica* [7]	*N. salina* [7]	*N. gaditana* [8]	*N. oceanica* [9]
A *	29625	428000		<25	220	<4						
B1	3.23	62.70		5.10-7.00	10.90	0.18						
B2	1.91	4.20		2.50-6.20	2.60	0.53					2.21	
B3	8.93	141				7.98					11	
B5	3.77					0.65						
B6	0.28	15.50		0.36-0.95	0.58	6.90						
B7	0.08			0.10-0.11	0.13							
B9 *	300			1700-2600	2000	0.02					2080	
B12 *	50.00	900		85-170	195	7.80					25	
C	19.10	49.80		100-320	300	79.20						
D2 *		1400		45.0	<35							
D3 *				35.0	<35							1–48
E	42.18	632	1980	18.0-35.0	7.00	20.28	210	48-233	2.12	4.41		
K *		2800										

* Values expressed as µg/100 g. [1] Fabregas and Herrero [52]; [2] De Roeck-Holtzhauer et al. [51]; [3] Santiago-Morales et al. [53]; [4] Brown et al. [54]; [5] Pereira et al. [16]; [6] Durmaz [56]; [7] Safafar et al. [57]; [8] Edelman et al. [55]; [9] Ljubic et al. [58].

As can be seen in Table 4, within individual studies, the vitamin content of different species can differ by a factor of ten or more. Even larger variations have been detected among studies [50]. For instance, the B12 content of *T. suecica* was documented by De Roeck-Holtzhauer et al. [51] to be 900 µg/100 g DW, whereas it was reported to be 50 by Fábregas and Herrero [52]. These marked differences must be attributed to variations in culture conditions, changes in the harvesting technique as well as specific analytical techniques employed by the different laboratories [54]. In this line, nutrient depletion enhanced the production of vitamins C and E in *T. suecica* [59]. Similarly, α-tocopherol accumulated along with N concentration decrease in *N. oculata*, or with the addition of

N and P in *T. suecica* [50]. Among four different microalgaes, *N. oceanica* it was the only one able to produce vitamin D3 and this production was enhanced by culture irradiation with ultraviolet B [58]. Pereira et al. [16] attributed the low values of several vitamins in *Tetraselmis* sp. CTP4 to the high temperatures (>50 °C) applied in the industrial spray-drying process to produce microalgae biomass.

2.6. Carotenoids

Carotenoids are a diverse and widespread class of bioactive compounds produced by microalgae which play essential roles as accessory light harvesting pigments and photoprotective agents. Carotenoids are divided into carotenes, oxygen-free hydrocarbons such as α-carotene or β-carotene, and xanthophylls, oxygenated derivatives of carotenes (lutein, violaxanthin, fucoxanthin and astaxanthin, among others).

Total content of carotenoids is extremely variable. In *Nannochloropsis*, the measured levels ranged from 0.4 mg/g DW [32] to 5.3 [57]. Banskota et al. [28] documented the highest levels of carotenoids in *T. chuii* (8.5 mg/g DW), although lower values were frequently found in the literature [37,59,60] for other strains of this genus. Table 5 summarizes the carotenoid profile of *Nannochloropsis* and *Tetraselmis* genera reported in the bibliography. Up to 15 different individual molecules were compiled from various studies. β-carotene was the only carotenoid detected in all samples, whereas violaxanthin is the xantophill present in most of the studies. A higher proportion of xantophills (up to 80%) than that of carotenes was usual, with a great quantitative and qualitative variety of compounds in the first group. As can be seen in Table 5, this composition showed large variations between different strains, too. For instance, the qualitative profile of *Tetraselmis* sp. Z3 was comparable to that of *Tetraselmis* sp. C6, but the contents of every molecule were significantly higher for the latter.

The whole variability denoted in Table 5 is justified by several pieces of evidence. Basic carotenoid synthesis pathways, mainly β-carotene, are similar in most microalgae. However, since the activation of carotenogenesis is under strict genetic control [60], different species or strains can accumulate unusual molecules via specific biosynthetic routes [61]. Additionally, as profusely observed for proteins and lipids, carotenoids synthesis by microalgae occurs in response to environmental stress factors, and culture conditions severely affect its production [15,60]. For example, in *Nannocholoropsis*, carotenoids can increase under low light irradiation or penetration, due to high culture density, restriction of P and S and changes in N-source ingredients in the culture media [17].

Table 5. Carotenoids composition (expressed as mg/g of dry weight, DW) of *Nannochloropsis* sp. and *Tetraselmis* sp. microalgae.

Compound	Carotenoids (mg/g DW)												
	T. suecica [1]	*T. suecica* [2]	*T. chuii* [2]	*T. chuii* [3]	*T.* sp. Z3 [4]	*T.* sp. C6 [4]	*N. gaditana* [1]	*N. gaditana* [5]	*N.* sp. BR2 [2]	*N. granulata* [3]	*N.* sp. [5]	*N. salina* [6]	*N. limnetica* [6]
Carotene													
α-carotene	0.41	0.20	0.17	1.70		0.02						0.08	
β-carotene	0.43	0.62	0.94	1.00	0.002	0.01	1.00		0.45	1.00		2.22	0.28
Lycopene				1.50									
Total Carotenes								1.15			1.16		
Xanthophyll													
Cantaxanthin				0.40						0.40		0.14	0.003
Zeaxanthin				0.10			0.01	0.90		0.10	0.20	0.58	0.14
Violaxanthin	0.82	1.41	0.54				3.36	4.45	1.08		4.17	1.68	1.23
Lutein	0.85	0.48	0.62	0.60	0.37	1.65		0.02					
Anteraxanthin		0.83	0.20						0.17				
Astaxanthin		2.26		0.10					0.32				
Fucoxanthin					0.05	0.08						0.01	0.18
Neoxanthin					0.12	0.79		0.01				0.05	0.42
Diadinoxanthin								1.44			1.18	0.14	
Vaucheriaxanthin												0.09	0.16
Alloxanthin												0.13	
Diatoxanthin													0.14

[1] Di Lena et al. [60]; [2] Ahmed et al. [62]; [3] Banskota et al. [28]; [4] Grubišic et al. [37]; [5] Ryckebosch et al. [63]; [6] Safafar et al. [57].

2.7. Phenolic Compounds

Phenolic compounds are a large chemical family of metabolites including acids, alcohols, and flavonoids among other molecules, which contain a polyphenol structure consisting of six-carbon aromatic rings. *Nannochloropsis* and *Tetraselmis* genera could be a potential source of phenolic compounds (Table 6) and, as it can be observed, their concentrations are highly variable.

Although the highest levels were detected in *N. gaditana* extract with 32.0 mg gallic acid equivalent (GAE)/g DW [64] followed by *T. suecica* with 28.03 mg GAE/g DW (Haoujar et al., 2019), most of the reported concentrations were below 10 mg GAE/g DW. In this regard, it should be taken into consideration that phenolic composition is strongly influenced by microalgae species and growth conditions [57,65]. Furthermore, the extracting solvents used (water, alcohols, or organic solvents) significantly affect phenolic content [66,67].

Table 6. Total content (mg of gallic acid equivalent/g of dry weight) and identified phenolic compounds in biomass of *Nannochloropsis* and *Tetraselmis* genera.

Microalgae	Total Content	Individual Phenolic Compounds	Ref.
Nannochloropsis sp.	1.4	—	[65]
Nannochloropsis sp.	0.6	Gallic, ferulic, protocatechuic, chlorogenic, hydroxybenzoic, syringic, vanillic	[67]
N. gaditana	32.0	—	[64]
	22.9	Caffeic, caffeoylglucoside, protocatechuic, *p*-coumaroyl tyrosine, quercetin, rhamnosyhexosyl-methyl-quercetin, apigenin-O-rutinoside, feruloylglucaricacid	[68]
N. granulata	6.0–8.0	—	[48]
N. limnetica	5.8	Gallic, caffeic, ferulic, cinnamic, salycilic	[57]
N. oculata	2.0	—	[65]
	4.1	—	[69]
N. salina	6.8	Gallic, 3,4 dihydroxy benzoic, ferulic, *p*-coumaric, salycilic	[57]
Tetraselmis sp.	3.7	—	[65]
Tetraselmis sp.	25.5	—	[64]
T. sp. IMP3	2.3	Gallic, gentisic, catechin hydrate, 4-hydroxybenzaldehyde, vanillic, caffeic, epicatechin, ferulic salicylic, naringenin-7-glucoside, luteolin-7-o-glucoside, rutin, ellagic, quercetin	[66]
T. sp. CTP4	0.76	Gallic, *p*-hydroxybenzoic, catechin hydrate, epicatechin	[66]
T. chuii	8.6	—	[69]
	20.0	—	[48]
T. suecica	1.7	—	[65]
	28.0	Caffeic, caffeoylglucoside, protocatechuic, dimethoxyflavone, *p*-coumaroyl tyrosine, apigenin-*O*-rutinoside, rhamnosyhexosyl-methyl-quercetin	[68]

There are few published studies regarding the identification and quantification of individual phenolic compounds in *Nannochloropsis* and *Tetraselmis* genera. Such works show that phenolic profiles may be quite different, even when comparing to microalgal species of the same genus (Table 6). For instance, the very significant differences detected by Cardoso et al. [66] between two *Tetraselmis* strains are noteworthy. Caffeic acid (53.3 mg/L) and luteolin-7-o-glucoside (9 mg/L) were the most abundant molecules in *Tetraselmis* sp. IMP3, but those compounds remained undetected in *Tetraselmis* sp. CTP4. Safafar et al. [57] only identified simple phenolic acids in *Nannochloropsis salina* and *Nannochloropsis limnetica* as gallic, 3,4-dihydroxy benzoic, caffeic, p-coumaric, salycilic, cinnamic and ferulic, the latter being the most abundant phenolic acid in both species (2.90 and 2.45 μg/g DW, respectively). In contrast, chlorogenic acid was clearly (489.5 μg/g DW) the richest phenolic acid in *Nannochloropsis* sp. [67]. Protocatechuic acid, caffeic acid, and p-coumaroyl tyrosine were detected in *T. suecica* and *N. gaditana*, with the highest content in the former [68]. For the flavonoids class, apigenin-O-rutinoside and rhamnosylhexosyl-methyl-quercetin were identified (Table 6).

3. Bioactivity of *Tetraselmis* sp. and *Nannochloropsis* sp. Microalgae

In addition to their nutritional role, foods are being recognized as a source of non-nutritive bioactive compounds able to prevent or reduce the risk of non-communicable disorders, such as neurodegenerative and cardiovascular diseases, obesity and diabetes, and cancer. Although plant foods are a food category containing both nutritive and bioactive compounds, the increase in the world population, subsequent food demand and the higher consumer knowledge on the food/diet relationship have promoted and engaged the search for new sustainable food sources by the food industry. Among these new sources, microalgae are becoming a promising functional food supply. The health benefits attributed to these aquatic microorganisms are associated to the presence of both primary metabolites or nutrients (lipids, proteins, carbohydrates, and vitamins) and secondary metabolites (phenolic compounds, carotenoids, among others). Biologically active substances from microalgae are capable of exhibiting antioxidant, antibacterial, antiviral, antitumor, regenerative, antihypertensive, neuroprotective and immunostimulating effects [70]. Thus, microalgae and derived compounds have an important potential as natural and sustainable dietary supplements for the prevention, management, and treatment of chronic disorders [71]. However, although the number of microalgae species was estimated to be over 168,000 [72], only few species are currently cultivated on a large scale and just a limited number have been studied for commercial purposes. Nevertheless, the biodiversity of microalgae would allow to increase the sources of bioactive metabolites with potential therapeutic applications as dietary supplements. In the following sections, the existing in vitro and in vivo evidence on the biological properties of compounds contained in *Tetraselmis* and *Nannochloropsis* sp. microalgae will be described.

3.1. *Antioxidant Activity of Microalgae and Derived Compounds*

Reactive oxygen species (ROS) and reactive nitrogen species (RNS) are produced under physiological conditions, also introduced into the body by environment pollution, unhealthy dietetic habits, smoking, lack of sleep, and irradiation. At low concentration, ROS and RNS play a vital role on cellular redox signaling and immune function. However, when their concentration increases or the antioxidant defenses decrease, the balance between them is broken, resulting in a state of oxidative stress. In this state, ROS and RNS provoke the oxidation of protein, lipid, and DNA leading to lipid membrane peroxidation, mitochondrial swelling and lysis, mutagenic actions, post-translational protein modifications, and finally cell death [73,74]. Thus, oxidative stress and free radicals are linked to the etiology and pathogenesis of multiple chronic diseases such as neurodegenerative and cardiovascular disorders, obesity, diabetes, several types of cancer, and accelerated aging [75].

The human body has an effective antioxidant defense system composed by enzymatic and non-enzymatic mechanisms. Still, these mechanisms are not complete, and to maintain the oxidative balance, additional antioxidants must be ingested through the daily diet. Thus, the research and consumption of natural antioxidants to prevent damage caused by oxidative stress and reduce the development of various chronic diseases has notably increased in the last years. These natural antioxidants are also attracting interest as natural preservatives to prevent oxidation and spoilage during food manufacture and storage [76]. Although most natural antioxidants currently available on the market are derived from terrestrial plants, microalgae are nowadays being considered as a potential source of natural antioxidant compounds by the food, cosmetic and nutraceutical industries [77].

In Table 7, the existing evidence on the antioxidant activity demonstrated for *Tetraselmis* and *Nannochloropsis* sp. microalgae extracts is summarized. Most studies reporting antioxidant activity have been conducted under in vitro conditions. The studies are mainly based on the radical [2,2′-azino-bis(3-ethylbenzothiazoline-6-sulfonic acid), ABTS], 2,2-diphenyl-1-picrylhydrazyl (DPPH), peroxyl, and hydroxyl scavenging capacity of microalgae extracts, although other mechanisms of action such as iron and copper chelating, ferric reducing antioxidant power (FRAP), and inhibition of lipid peroxidation have also been demonstrated. In general, the results are very heterogeneous depending on the species and strains studied, the antioxidant activity assays, the extraction conditions and the concentrations used. Moreover, the variety in the biochemical composition and the capacity of microalgae to produce a broad range of antioxidant compounds is the basis for the diversity in their antiradical activity. Although few associations have been made between the antioxidant effects of microalgae and the metabolites responsible for the observed effects, some correlations have been detected with phenolic compounds including flavonoids, carotenoids, PUFAs and protein/peptide fractions [77]. As in plants, phenolic compounds are one of the main components responsible for the antioxidant defense mechanisms in microalgae. These compounds mainly act to neutralize ROS through hydrogen atom transfer, although other mechanisms have been described, such as inactivation of radicals by monoelectronic transfer and chelation of transition metals involved in the Fenton reaction, thus preventing the formation of highly reactive hydroxyl radicals [78].

Table 7. Mechanisms of antioxidant activity and potential responsible compounds of the observed effects demonstrated for *Tetraselmis* sp. and *Nannochloropsis* sp. microalgae extracts by biochemical assays and cell models.

Microalgae	Pretreatment/Extraction Conditions	Mechanisms	Responsible Compound	Ref.
Tetraselmis sp.	(1) Hexane (2) Sequential extraction with ether, acetone and water	DPPH radical scavenging, and iron and copper chelating	PUFAs and phenolic compounds	[79]
	Ethanol and ethanol/water or ultrasound + water	DPPH radical scavenging	—	[64]
	Methanol/dichloromethane + ultrasound or acetone + ultrasound	ORAC radical scavenging, and inhibition of lipid peroxidation (TBARS)	Xanthophylls	[80]
	Ethanol/water or hexane + ethyl acetate + hot water	ABTS radical scavenging, FRAP, and inhibition of linoleic acid oxidation	Phenolic compounds	[65]
	Methanol/dichloromethane + ultrasound	Inhibition of lipid peroxidation (TBARS)	—	[77]
T. sp. KCTC12236BP and KCTC12432BP	(1) 1 N H_2SO_4, 1 M NaOH and distilled water (2) Ethanol precipitation	ABTS and DPPH radical scavenging and FRAP	Polysaccharides	[81]
T. sp. KCTC12236BP	Ultrasound + water:ethanol	DPPH radical scavenging	Total phenolic compounds and epigallocatechin gallate	[82]
T. sp. M8	Water, hexane or ethyl acetate	ORAC	Carotenoids	[62]
T. sp. IMP3, and *T.* sp. CTP4	Water or ethanol 96%	ABTS and DPPH radical scavenging and FRAP	Phenolic compounds and PUFAs	[66]
T. chuii	Methanol or (1) hexane/dichloromethane and (2) acetone/water/acetic acid	ORAC and DPPH radical scavenging	PUFAs and phenolic compounds	[28]
	Dichloromethane (Folch method)	ABTS and DPPH radical scavenging	PUFAs and other compounds	[76]
	(1) Pulsed electric field (2) DMSO or distilled water	ABTS radical scavenging	Polyphenols and pigments	[83]
	Ethanol, ethyl acetate, or water	DPPH radical scavenging, and iron and cupper chelating	—	[84]
	Water, hexane or ethyl acetate	ORAC	Carotenoids	[62]
T. marina IMA043	(1) Ultrasound (2) Water, methanol or dichloromethane	DPPH and ABTS radical scavenging, iron/cupper chelating, and FRAP	SFA, MUFAs and phenolic compounds	[85]
T. suecica	Distilled water, hexane, heptane, ethyl acetate, acetone, ethanol or methanol	ABTS radical scavenging, and FRAP	Lutein	[86]
	Ultrasound + methanol:water, methanol or ethanol	DPPH and ABTS radical scavenging, and β-carotene bleaching	Multiple compounds	[87]
	Ethanol:water	DPPH radical scavenging	Carotenoids	[88]
	Ethanol/water or hexane + ethyl acetate + hot water	ABTS radical scavenging, FRAP, and inhibition of linoleic acid oxidation	Phenolic compounds	[65]
	Water, hexane or ethyl acetate	ORAC	Carotenoids	[62]
	Water + ultrasound + acetone	Cellular antioxidant and lipid peroxidation inhibition in HepG2 cells	—	[89]
T. suecica TES2	(1) Ultrasound + chloroform:methanol or ethanol (2) Enzymatic hydrolysis	ORAC, DPPH, and ABTS radical scavenging	Protein/peptide fractions	[90]
T. suecica (Kylin) Butcher	(1) Phenolics removal with polyvinylpyrrolidone (2) Ethanol or N-cetylpyridinium bromide extraction	ABTS and DPPH radical scavenging	Exopolysaccharides	[91]
T. striata CTP4	Ethanol, ethyl acetate, or water	DPPH radical scavenging, and iron and cupper chelating	—	[84]
T. tetrathele	Methanol	DPPH radical scavenging, and inhibition of lipid and linoleic acid peroxidation	—	[92]

Table 7. *Cont.*

Microalgae	Pretreatment/Extraction Conditions	Mechanisms	Responsible Compound	Ref.
Nannochloropsis sp.	Ethyl acetate, methanol, water, isopropanol or petroleum ether	Total capacity, hydroxyl and NO radical scavenging, metal chelating, and FRAP	Fatty acids (hexadecanoic acid)	[93]
	Ethanol:water	Inhibition of LDL oxidation and ROS inhibition in H_2O_2-induced macrophages J-774A	Lyso-diacetylglyceryltrimethylhomoserine	[94]
	Methanol	Total capacity; DPPH, superoxide, and hydroxyl radical scavenging	—	[95]
	Ethanol, ethyl acetate, or water	DPPH radical scavenging, and iron and cupper chelating	—	[84]
	Methanol + ultrasound	ABTS and DPPH radical scavenging, FRAP, and ferrous ion-chelating ability	Phenolic compounds and carotenoids	[57]
	Water, hexane or ethyl acetate	ORAC	Carotenoids	[62]
	Hexane, dichloromethane, chloroform or methanol	DPPH, superoxide anion radical scavenging, FRAP, and ferrous-ion chelating	—	[96]
	Ethanol + water	ABTS radical scavenging	—	[97]
N. sp. NNX1	(1) Ultrasound + chloroform:methanol or ethanol (2) Enzymatic hydrolysis	ORAC, DPPH, and ABTS radical scavenging	Protein/peptide fractions	[90]
N. sp. SBL1 and SBL4	Methanol	DPPH radical scavenging and iron and cupper chelating	Phenolic compounds	[98]
N. gaditana	Ethanol and ethanol/water or ultrasound + water	DPPH radical scavenging	—	[64]
	Ethanol:water:12N HCl	ABTS and DPPH radical scavenging, and lipid peroxidation inhibition	—	[99]
	(1) High pressure disruption (2) Alkaline pH and isoelectric protein precipitation (3) Papain hydrolysis	ORAC radical scavenging	Protein/peptide fractions	[100]
	Supercritical CO_2	DPPH radical scavenging, β-carotene bleaching, and FRAP	Lipids	[101]
N. granulata	Methanol or (1) hexane/dichloromethane and (2) acetone/water/acetic acid	ORAC and DPPH radical scavenging	PUFAs and phenolic compounds	[28]
N. limnetica 0065NA	Chloroform:methanol	DPPH radical scavenging	Polar lipids	[35]
N. oceanica	Dichloromethane (Folch method)	ABTS and DPPH radical scavenging	PUFAs and other compounds	[76]
N. oceanica 0011NN	Chloroform:methanol	DPPH radical scavenging	Polar lipids	[35]
N. oculata	Ethyl acetate or methanol	DPPH and NO radical scavenging, metal chelating, and FRAP	Phenolic and flavonoid compounds	[102]
	(1) Supercritical CO_2 (2) Subcritical n-butane extracts	DPPH radical scavenging	Carotenoids	[103]
	(1) Methanol (2) Water	DPPH radical scavenging	Polysaccharides	[104]
	Methanol, ethanol:water, chloroform, or hexane (Soxhlet)	DPPH radical scavenging	Hexanedioic acid, bis (2-ethylhexyl)ester	[105]
	Methanol:water:acetic acid:ascorbic acid + ultrasound or methanol:acetone:hexane or ethyl acetate:hexane + ultrasound or methanol:water:HCl + ultrasound	ABTS and ORAC radical scavenging, and FRAP	Polyphenols, carotenoids, chlorophylls, and triterpenoids	[2]
	Methanol + ultrasound	DPPH, H_2O_2, and ABTS radical scavenging	—	[106]
	Ethanol/water or hexane + ethyl acetate + hot water	ABTS radical scavenging, FRAP, and inhibition of linoleic acid oxidation	Phenolic compounds	[65]
	Hexane or methanol	DPPH radical scavenging, and iron and copper chelating	Phenolic compounds and PUFAs	[69]
	Methanol	DPPH radical scavenging, and inhibition of lipid and linoleic acid peroxidation	—	[92]

ABTS: 2,2′-azino-bis(3-ethylbenzothiazoline-6-sulfonic acid); DMSO: dimethyl sulfoxide; DPPH: 2,2-diphenyl-1-picrylhydrazyl; FRAP: ferric reducing antioxidant power; MUFA: monounsaturated fatty acids; NO: nitric oxide; ORAC: oxygen radical absorbance capacity; PUFA: polyunsaturated fatty acids; ROS: reactive oxygen species; SFA: saturated fatty acids; TBARS: thiobarbituric acid reactive substances.

Total phenolic compounds present in extracts obtained from *Tetraselmis* sp., *T. suecica* and *N. oculata* with ethanol/water or hexane/ethyl acetate/hot water mixtures were demonstrated to scavenge ABTS radicals, inhibit lipid peroxidation and show FRAP [65]. These compounds have also been reported to mostly contribute to the DPPH radical scavenging capacity of *Tetraselmis* sp. KCTC12236BP extracts [82], the DPPH radical scavenging and iron and cupper chelating capacities of *Nannochloropsis* sp. SBL1 and SBL4 methanolic extracts [98], and the DPPH and NO radical scavenging activity, metal chelating ability, and FRAP values of *N. oculata* extracts [102]. In a recent work, phenolic compounds are recognized as the major factor responsible for the ABTS and DPPH radical scavenging activity and the FRAP values of extracts obtained from *T. marina* IMA043 strain, although the authors suggest that the remaining activity that cannot be explained by the presence of phenols could be ascribed to other molecules present in the extract [85]. Similar findings had been reported in previous studies that recognized both phenolic compounds and PUFAs as the contributors on the antioxidant activity exerted by extracts from *N. oculata* [69], *Tetraselmis* sp. [79], *T. chuii* and *N. granulata* [28]. The correlation between PUFAs content and antioxidant activity of microalgae has been recently reported [76]. These authors determined that species (*C. vulgaris*, *Chlorococcum amblystomatis*, *Scenedesmus obliquus*, and *Phaeodactylum tricornutum*) with higher abundance of PUFAs than *N. oceanica* showed higher DPPH and ABTS radical scavenging activity, while neither omega-3/omega-6 ratio nor MUFAs content exerted any influence. Banskota et al. [28] also suggested that the higher the unsaturation level (especially PUFA content), the better the lipophilic (L)-ORAC values of microalgae extracts. Thus, *T. chuii* had the lowest PUFAs and MUFAs content, and also low L-ORAC values. However, the relationship between microalgae lipids and antioxidant activity needs to be completely elucidated.

Carotenoids have been used as feed for aquaculture and food colorants, becoming very popular as dietary supplements in the last years. They are potent antioxidants for human health and are involved in the modulation of body functions such as cellular signaling and gene expression [107]. Their ability to react with free radicals through hydrogen atom transfer, monoelectronic transfer and adduct formation is already well known [108]. Moreover, due to their conjugated double bond system, carotenoids play an important role in quenching ROS released during photosynthesis, especially singlet oxygen [65]. The carotenoids-rich extracts obtained from different microalgae species have demonstrated to exert potent antioxidant effects through multiple mechanisms. Therefore, the ethanol/water extract from *T. suecica*, containing high levels of carotenoids, showed marked and dose-dependent DPPH radical scavenging activity, comparable to that shown by known antioxidants such as α -tocopherol and ascorbic acid [88]. In this study, the protective effects of the extract were also confirmed in H_2O_2-induced lung A549 cells, finding that it could revert the cell viability inhibitory effects of the chemical oxidant, reaching up to 100% recovery at the highest extract dose. Moreover, the extract targeted the expression of dehydrocholesterol reductase-24 (DHCR24) and prostaglandin reductase 2 (PTGR2) genes and proteins, and reduced the levels of prostaglandin E2 (PGE2) [88]. The study by Feller and coworkers [103] also reported that carotenoid compounds were directly correlated with the DPPH radical scavenging capacity of *N. oculata* extracts obtained by both supercritical CO_2 and subcritical n-butane extraction methods, independent of the extraction method tested. One of the major carotenoids produced by this microalgae specie is zeaxanthin; thus, future studies focusing on demonstrating its contribution on the antioxidant effects of extracts as well as on optimizing the extraction conditions to reach higher carotenoids yield should be of interest. The total carotenoid and lutein contents were also strongly correlated with the ORAC values measured in extracts obtained from different microalgae species including *Tetraselmis* sp., while no correlation was detected for the β-carotene [77]. Moreover, other compounds present in the extracts such as phenolic compounds and fatty acids were suggested to contribute to the lipid peroxidation inhibition shown by the microalgae species. Lutein is a fat-soluble carotenoid, known by its health benefits as antioxidant, anti-inflammatory and protective agent against cardiovascular and Alzheimer's diseases [109]. In a recent study, the solvent extraction conditions were optimized by response surface methodology (RSM) to improve lutein recovery from microalga *T. suecica* [86]. At the optimized conditions, the lutein content was more than 3-fold higher than that of the control group (before optimization), and the extract showed higher ABTS radical scavenging activity.

In the last years, other compounds present in microalgae have been determined to contribute to the health benefits, namely it was discovered that microalgae proteins can exert biological activities through peptides included into their sequence and released by use of exogenous enzymes from microbial or mammalian origin as well as via gastrointestinal proteases [14]. Bioactive peptides have been determined to exert multiple bioactivities such as antioxidant, antimicrobial, anti-hypertensive, chemopreventive, and modulatory of immune, nervous and/or metabolic systems. However, up to date, few data are available on the production of these bioactive peptides from *Tetraselmis* or *Nannochloropis* proteins. Medina et al. [100] reported that the *N. gaditana* microalgal biomass hydrolyzates with papain showed potent antioxidant activity measured by the ORAC assay. Norzagaray-Valenzuela et al. [90] analyzed the antioxidant capacity of microalgae residual biomass and protein hydrolyzates with alcalase of three green microalgae species: *T. suecica* TES2, *Nannochloropsis* sp. NNX1, and *Dunaliella tertiolecta*. The hydrolyzates showed higher ORAC, DPPH, and ABTS radical scavenging activities than their biomass at the assayed concentrations. Moreover, hydrolyzates from *Nannochloropsis* sp. showed anti-aging effects through inhibition of elastase and hyaluronidase. Although limited, these preliminary findings suggest the role of microalgae proteins as source of multifunctional peptides with promising applications in food, pharmaceutical, and cosmetical industries.

Multiple biological effects such as antioxidant, immunomodulatory antimicrobial, and antitumoral have been ascribed to polysaccharides [110]. Regarding the antioxidant activity, microalgae polysaccharides are known free radical scavengers, and therefore have antioxidant effects preventing oxidative damage in living organisms. Kashif et al. [81] purified water-soluble polysaccharides from defatted microalgal biomass of *Tetraselmis* sp. KCTC 12236BP and KCTC 12432BP, finding a strong correlation between these compounds and the antioxidant, antifungal, and tyrosinase inhibitory activities. More recently, the adaptation of autotrophic to heterotrophic culture was carried out with *T. suecica* (Kylin) Butcher to increase the production of exopolysaccharides and their ABTS radical scavenging capacity and cytotoxic effects against human leukemia, breast, and lung cancer cells [91].

Even though cell models provide more physiological information by considering the bioavailability and metabolism of tested compounds, the available data about the antioxidant activity of microalgae extracts using cellular assays are still scarce (Table 7). In the study of Lauritano et al. [89], the antioxidant activity of the crude extracts of 32 microalgal species including *T. suecica* was evaluated using two different assays in hepatocellular liver carcinoma HepG2 cells: the cellular lipid peroxidation antioxidant activity and the cellular antioxidant activity assays. In addition, the anti-inflammatory, antibacterial, anticancer, and antidiabetic properties of the extracts were evaluated, concluding that the production of primary and secondary bioactive metabolites in microalgae is influenced by the species and strain and the culture media and conditions, growth phases, and clones. This metabolic flexibility favors the discovery of novel bioactive compounds for the management/control of human diseases. Khatib et al. [94] examined the antiatherogenic effects of *Nannochloropsis* sp. extracts, reporting that the 70% ethanol–water extract inhibited macrophages J-774A and LDL oxidation, increased protein paraoxonase 1 (PON1) activity, and protected it from the deleterious effects of linoleic acid hydroperoxide. The authors isolated and purified the compound responsible for the observed effects, the structure of which was elucidated as the lyso-diacylglyceryltrimethylhomoserine lipid.

As it has been described for cellular assays, the animal models reported to evaluate the antioxidant activity of microalgae extracts are still limited in comparison to in vitro assays. In the case of *Tetraselmis* sp. and *Nannochloropsis* sp., few studies can be found in the literature, and they are mainly focused on evaluating the impact of microalgae meal on the antioxidant status of shrimp or fish, suggesting their potential as an alternative to fishmeal in aquafeed [111,112]. Recently, male streptozotocin-induced diabetic rats were fed control either diet supplemented with 10% *N. gaditana* or not, for 2 months. The administration of microalgae was effective in attenuating oxidative stress and inflammation in diabetic rats, reducing glucose and glycated hemoglobin levels, and improving the renal and hepatic functions [113]. Thus, although the number of studies is still scarce, it seems that the inclusion of microalgae directly into the diet could exert benefits on the animal physiology, indicating the promising potential of their use in human or animal nutrition as functional ingredients. However, the digestibility of the microalgae and bioavailability of the antioxidant compounds into the food/supplement matrices are research areas that need to be explored.

3.2. Antimicrobial Activity of Tetraselmis and Nannochloropsis Microalgae

In the last years, the demand for new and effective antimicrobial compounds has notably increased due to the evolution of microbial pathogens and the consequent antibiotic resistance. Among the natural sources of antimicrobials, microalgae are being considered one of the most promising because of their capacity to produce numerous and diverse bioactive metabolites that allow them to defend themselves against pathogenic bacteria, viruses, and fungi in the aquatic environment [114]. The antimicrobial activity of different microalgae extracts has been mainly associated to fatty acids [115], phenolic compounds [116], and pigments [117]. Maadane et al. [118] evaluated the antimicrobial effects of the ethanolic extracts obtained from nine microalgae species including *N. gaditana* and *Tetraselmis* sp. against bacteria *Escherichia coli, Pseudomonas aeruginosa* and *Staphylococcus aureus*, the yeast *Candida albicans* and the fungus *Aspergillus niger*. The authors linked the observed antimicrobial effects to the contents of the extracts in fatty acids, carotenoids, and phenolic compounds. In the recent study of Wali et al. [106], the extracts obtained from *N. oculata* by using methanol under sonication showed potent inhibitory activity against bacteria *S. aureus* ATCC 25923, *P. aeruginosa* ATCC 27853, and *E. coli* ATCC 25922, and moderate effects against the yeast *C. albicans* ATCC 24433. The extracts also showed potent antioxidant and anticancer properties on breast cancer MDA-MB-231 cells. The terpenoids, carotenoids, polyphenolic, and fatty acids contained in the extracts were suggested as the major factor responsible for the observed effects. More recently, the ethanolic extracts of *N. oceanica* were determined to exert inhibitory effects against *Vibrio harveyi* in an in vitro co-culture assay, although the responsible compounds were not identified [119]. The growth inhibitory effects on *Vibrio anguillariim*, and *Vibrio salmonicida* reported for extracts obtained from *T. suecica* were not associated with any specific compound [120]. Vibriosis is associated with fish infections and death, resulting in enormous economic loss for farmers. Thus, the antibacterial properties of microalgae might be exploited as an alternative for disease prevention and management in the aquaculture industry [121].

In addition to phenolic compounds and fatty acids, peptides and polysaccharides have also been determined to contribute on the antimicrobial activity demonstrated for microalgae extracts. Three peptide sequences identified in the acid extracts of *T. suecica* were determined to exert potent antimicrobial activity against both Gram-positive and Gram-negative bacterial strains [122]. In this work, substitutions of leucine, tyrosine, threonine, and methionine by lysine resulted in a notable increase in the bioactivity, particularly against Gram-positive bacteria. This could be due to the increase in the peptide's positive net charge, correlated with the lysine side chain, which is able to interact with the negatively charged bacterial membranes [123]. Water-soluble polysaccharides extracted from *N. oculata* and *Tetraselmis* sp. were determined to inhibit the growth of both bacteria and yeasts [81,104]. Although the mechanisms involved in their antimicrobial activity have not been completely elucidated, the capacity of polysaccharides is well known to impact the cytoplasm permeability and to participate in the DNA decomposition after polysaccharide/DNA binding, as well as affect the denaturation of essential bacterial proteins [124]. Therefore, the presence of different antimicrobial compounds in microalgae and the potential synergistic action between them opens the door to new applications of these organisms in aquaculture as well as in the nutritional and human medical fields.

3.3. Anti-Carcinogenic Effects of Tetraselmis and Nannochloropsis sp. Microalgae Compounds

Cancer has become one of the world's leading causes of disease and death with important social and economic repercussions. Thus, the current interest is focused on the search of novel sources of bioactive compounds able to act on carcinogenesis-associated biomarkers and pathways. Among them, natural compounds obtained from various sources have been demonstrated to exert anticancer potential, therefore being promising candidates for the development of new drugs. As a result, currently, the percentage of drugs derived from natural sources corresponds to approximately 60% [125].

Several research studies reported microalgae species including *Tetraselmis* sp. and *Nannochloropsis* sp. with anti-cancer properties (Table 8). Some of these works evaluated the cytotoxic effects of a single application of microalgal extracts. Chloroform extracts obtained from *T. suecica* reduced the viability of breast cancer MCF-7 and 4T1 cells [126]. Moreover, potentiation of the cytotoxic effects and induction of apoptosis were reported when microalga extracts were co-applied with silver nanoparticles [105] or tamoxifen [127]. Methanolic extracts from *N.*

oculata were also determined to inhibit breast cancer MDA-MB-231 cell viability in a dose- and time-dependent manner in addition to exerting antioxidant and antimicrobial properties [106]. Different compounds present in microalgae extracts have been described as major contributors to the anticancer properties. Thus, a sterol-rich fraction of cultured marine microalga *N. oculata* showed marked cytotoxic and apoptosis inducing effects on leukemia HL-60 cells, and anti-inflammatory activities against lipopolysaccharide-stimulated RAW 264.7 macrophages [128]. Lipids, mainly omega-3 PUFAs, were determined to be the main factor responsible for the cytotoxic effects against colon cancer HCT-116 cells of *N. gaditana* extracts obtained by an approach combining enzyme and ultrasounds [129]. A water-soluble polysaccharide-enriched extract from *N. oculata* was reported to inhibit proliferation of HeLa cells in a dose-dependent manner, in addition to being antimicrobial and anticholinesterasic [104].

Table 8. Biological activities demonstrated for *Tetraselmis* and *Nannochloropsis* sp. microalgae extracts.

Microalgae Specie/Strain	Antimicrobial Activity (Microorganism)	Anti-Carcinogenic Activity (Cell Model)	Other Bioactivities	Ref.
Tetraselmis sp.	—	Cell viability inhibition (HepG2)	Acetylcholinesterase inhibition	[79]
T. sp. KCTC12236BP and KCTC12432BP	Antifungal (*Candida albicans* and *Penicillium italicum*)	—	Tyrosinase inhibition	[94]
T. chuii	—	Cell viability inhibition (HepG2)	Calcium chelating	[84]
T. suecica	—	Cell viability inhibition and apoptosis induction (MCF-7 and 4T1)	—	[127]
	Antibacterial (*Escherichia coli*, *Pseudomonas aeruginosa*, and *Staphylococcus aureus*) Antifungal (*Candida albicans* and *Aspergillus niger*)	—	—	[118]
	—	Anti-angiogenic through over-expression of VEGF (PC-3 cells)	—	[87]
	—	Repairing effects through increasing expression of DHCR24 and PTGR1 genes and proteins and reducing PGE2 (A549)	Repairing effects on reconstructed human epidermal tissue cells (EpiDerm™)	[88]
T. suecica CCAP904	Antibacterial (*Escherichia coli*, *Salmonella typhimurium*, *Pseudomonas aeruginosa*, *Bacillus cereus*, *Staphylococcus aureus*, *Listeria monocytogenes*, *Micrococcus luteus*)	—	—	[122]
T. suecica TES2	—	—	Anti-aging potential through inhibition of elastase and hyaluronidase	[90]
T. suecica (Kylin) Butcher	—	Cell viability inhibition (MCF-7, HL-60, and NCI-H460)	—	[91]
T. striata CTP4	—	—	Calcium chelating Anti-inflammatory through decreasing TNF-α in LPS-induced macrophages THP-1	[84]
Nannochloropsis sp.	—	—	Increase in anti-atherogenic paraoxonase 1 activity	[94]
	—	—	Calcium chelating Anti-inflammatory through decreasing TNF-α in LPS-induced macrophages THP-1	[84]
N. sp. NNX1	—	—	Anti-aging potential through inhibition of elastase and hyaluronidase	[90]
N. gaditana	—	Cell viability inhibition (Caco-2 and HepG2)	—	[130]
	—	Cell viability inhibition (HCT-116)	—	[129]
	—	—	Skin protection properties by mediating oxidative responses and apoptosis (H_2O_2-challenged dermal fibroblasts)	[131]
N. oceanica	Antibacterial (*Vibrio harveyi*)	—	—	[119]
N. oculata	Antibacterial (*Enterococcus faecalis*, *Staphylococcus aureus*, *Escherichia coli*, and *Pseudomonas aeruginosa*) Antifungal (*Candida albicans*, *Candida glabrata*, *Candida kreusei*, and *Candida parapsilosis*)	Cell proliferation inhibition (HeLa)	Anti-cholinesterase activity	[104]
	—	Cell viability inhibition and apoptosis induction (MCF-7 and 4T1)	—	[105]
	—	Cell viability inhibition and apoptosis induction (MCF-7 and 4T1)	—	[127]
	—	Cell viability inhibition and apoptosis induction (HL-60)	Anti-inflammatory through decreasing LPS-induced iNOS and COX-2 protein levels (RAW264.7)	[128]
	—	—	Anti-aging through inhibition of acetylcholinesterase and butyrylcholinesterase Anti-diabetic through inhibition of α-amylase Anti-obesity through inhibition of pancreatic lipase	[2]
	Antibacterial (*Staphylococcus aureus*, *Pseudomonas aeruginosa*, and *Escherichia coli*) Anti-fungal (*Candida albicans*)	Cell viability inhibition (MDA-MB-231)	—	[106]

COX-2: cyclooxygenase-2; DHCR24: dehydrocholesterol reductase-24; iNOS: inducible nitric oxide synthase; PGE2: prostaglandin E2; PTGR1: prostaglandin reductase 1; TNF-α: tumor necrosis factor-alpha; LPS: lipopolysaccharide; VEGF: vascular endothelial growth factor.

Multifunctionality was also reported for exopolysaccharides purified from the heterotrophic culture of *T. suecica* (Kylin) Butcher that were determined to exert ABTS radical scavenging activity and cell viability inhibitory properties against leukemia, breast, and lung cancer cells [91]. The multifunctionality could represent an additional value for these microalgal products. Proteins have also been described as contributors to the cytotoxic effects of microalgae extracts. Thus, among 488 proteins identified by proteomic analysis of *N. gaditana*, Carrasco-Reinado et al. [130] selected the UCA01 protein belonging to the prohibitin family to evaluate its bioactivity. This protein was demonstrated to exert potent antiproliferative activity against colon adenocarcinoma Caco-2 and hepatocellular carcinoma HepG2 cells, without affecting nontumorigenic cells. Therefore, the recent findings on the anticancer properties of *Tetraselmis* sp. and *Nannochloropsis* sp. microalgae open new possibilities to produce bioactive ingredients with potential health benefits and applications in the food and nutraceutical industry.

4. Digestibility of *Tetraselmis* sp. and *Nannochloropsis* sp. Microalgae Biomass

In vitro models that simulate human gastrointestinal digestion provide valuable information on the bioavailability of potentially bioactive compounds present in microalgae biomass. In general terms, the disruption of the cell wall of microalgae prior to digestion is highly recommended to facilitate the action of digestive enzymes. This should be considered especially for microalgae with thick and rigid cell walls, such as *Nannochloropsis* sp., as their digestibility is affected by the selected disruption treatment [14].

In vitro crude protein digestibility (IVPD) is often used to assess the digestibility of microalgae biomass and to determine its potential as a protein source. Wild et al. [22] reported that IVPD was in the range of 48–59% for *N. oceanica* and *N. oculata* when the cell wall was not disrupted, and it was increased to 78–80% when microalgae biomass was treated with a stirred ball mill [22]. Lipid-extracted *N. granulata* and *T. chuii* showed significantly higher levels of IVPD (>85%) when compared to the same species of whole microalgae [48]. On the other hand, in vitro lipid digestibility (IVLD) is determined by the comparison of free fatty acid concentration released after digestion with the total fatty acid content of the original sample. In this line, lipid digestion of *Nannochloropsis* sp. was considered incomplete in untreated biomass, accounting for 36–40% of IVLD. The application of high pressure homogenization in these microalgae suspensions substantially modified IVLD (56–62%) and it also increased the bioaccessibility of omega-3 PUFAs and carotenoids during the small intestinal phase [132]. Recent research comparing different species of microalgae reported lower IVLD values for *N. gaditana* treated by bead milling (33.3%), suggesting that such pretreatment was not sufficient for effective cell disruption [27].

In terms of in vitro carbohydrate digestibility (IVCD) in different species of microalgae, their values tend to mimic those of organic matter digestibility except in microalgae with high ash content (>20%) [30]. Free glucose is the main product of IVCD, and its release depends on microalgae characteristics. Verspreet et al. [27] observed that disrupted *N. gaditana* only yielded 1.8 g of free glucose/100 g of dry matter, while other species produced 10 times more. They related this fact to the β-linkages of glucose units that would not be susceptible to degradation by human α-amylases or disaccharidases [27]. In addition, it is also important to consider that culture conditions can greatly affect the carbohydrate content of microalgae. For instance, *T. suecica* presents an average carbohydrate content of approximately 10% when it is grown in a medium that provides all the nutrients, and it can reach 37% when growing medium is deprived of a nitrogen source. However, it appears that those changes in carbohydrate composition would have little effect on IVCD and organic matter digestibility [30].

In vivo investigations designed to understand the human digestion process of microalgae biomass are very scarce and further research is needed. Functional extracts obtained from untreated and treated *N. gaditana*, through a combined thermal and high-pressure treatment, were recently evaluated in vivo using a rat model with a 20% dietary inclusion level of microalgae [99]. Although both untreated and treated microalgae showed interesting potential for dry matter digestibility and antioxidant capacity (i.e., polyphenol content, ABTS and inhibition of lipid peroxidation), treated *N. gaditana* was able to improve all biomarkers. The in vivo apparent digestibility coefficient accounted for 67% and 83% in rats fed the untreated and treated microalgae biomass, respectively. In addition, the fecal microstructure of rats fed untreated *N. gaditana* clearly showed undigested microalgae and their microbiota was different compared to that of the rats fed the treated microalgae. It would suggest that the technological treatment not only favors the release of certain metabolites but could also promote the growth of specific microbiota [99]. Overall, we can conclude that microalgae biomass processing leading to cell disruption is a key factor for nutrient release and consequent digestibility of the compounds of interest.

5. Digestibility of *Tetraselmis* sp. and *Nannochloropsis* sp. in Food Products

T. chuii dried biomass has been authorized in the EU for commercialization as novel food and food supplement [7,133]. The intake of this microalgae is safe, and its unique nutritional composition makes it a promising sustainable ingredient for the agrifood industry [134,135]. *T. chuii* has been used to replace wheat flour in the manufacture of protein-enriched bread [20] and it has also been incorporated in buckwheat, rice flour and potato starch formulations to increase the bioactivity of gluten-free bread [136]. Qazi et al. [137] evaluated the IVPD of raw and ethanol-treated *T. chuii* biomass, either alone or incorporated into wheat bread, following the standardized INFOGEST static in vitro digestion model. Breads including 12% of disrupted *T. chuii* showed significantly higher enzymatic degradation of proteins into small peptides than bread without microalgae. Dissolved protein represented 83% in the commodity wheat flour bread, and it was reduced to 59–67% by incorporating microalgae biomass. As a result, IVPD decreased in breads formulated with *T. chuii*, especially when microalgae biomass was treated with ethanol [137].

In vitro digestibility of *T. suecica* in food products has also been evaluated. Crackers formulated with 2 and 6% of untreated microalgae biomass showed a similar dry matter digestibility to crackers without microalgae (~80–85%). IVPD values ranged from 50 to 65% in *T. suecica* crackers, slightly lower than the 75% IVPD of commercial crackers, but the level of added microalgae biomass did not affect either dry matter digestibility or IVPD [138]. Similar contents of *T. suecica* were also incorporated into cookies when wheat flour was replaced by 2 and 6% of untreated microalgae biomass. No changes were observed when comparing the in vitro digestibility of cookies with distinct species of microalgae at different concentrations. However, prior to cookies manufacture, the evaluation of the in vitro digestibility revealed differences between the microalgae species [139]. This would highlight the importance of the food matrix in in vitro digestive simulations and thus, further research is encouraged to fully understand the digestibility of microalgae in complex food products. On the other hand, to our knowledge, there is no published research evaluating the digestibility of *Nannochloropsis* sp. incorporated into a food matrix. Although there is a huge commercial interest in this microalga as a food ingredient, it has not yet been authorized as novel food due to the difficulty of complying with all strict safety criteria imposed by the current legislation [17]. In the coming years, an increase in patent applications and commercial authorizations related to *Nannochloropsis* sp. is expected, but it is essential to evaluate the digestibility of these microalgae-enriched food products before their launch on the market.

6. Conclusions

Tetraselmis sp. and *Nannochloropsis* sp. are excellent sources of a diversity of bioactive compounds, including lipids, proteins, carbohydrates, minerals, and vitamins. Their variable composition depends to a large extent on the growing and drying conditions of the microalgae biomass, which substantially affects their bioactivity and digestibility. Most of the research on bioactivity has been focused on determining the antioxidant capacity of *Tetraselmis* and *Nannochloropsis* sp. extracts under different pretreatment and extraction conditions. Research on other potential bioactivities (i.e., antibacterial, antifungal, anti-inflammatory, anti-diabetic, anti-obesity, anti-aging, anti-cancer) is expected to increase in the coming years to define the key compounds responsible of such bioactivities and design microalgae culture conditions that allow to maximize their potential. Regarding the digestibility of *Tetraselmis* and *Nannochloropsis* microalgae, the available information is still scarce. Further investigations are encouraged to assess the potential impact that the food matrix could exert on the behavior of the microalgae under physiological conditions to select the most suitable form of microalgae administration.

Author Contributions: Writing—original draft preparation, S.P., M.A.d.l.F., P.G.-C., B.H.-L.; writing—review and editing, M.A.d.l.F., P.G.-C., B.H.-L.; funding acquisition, P.G.-C., B.H.-L. All authors have read and agreed to the published version of the manuscript.

Funding: This research was funded by the Spanish Ministry of Science and Innovation, grant number PID2021-122989OB-I00 and the Spanish National Research Council, grant number 202270I067.

Institutional Review Board Statement: Not applicable.

Informed Consent Statement: Not applicable.

Data Availability Statement: Not applicable.

Conflicts of Interest: The authors declare no conflict of interest.

References

1. Barros de Medeiros, V.P.; da Costa, W.K.A.; da Silva, R.T.; Pimentel, T.C.; Magnani, M. Microalgae as source of functional ingredients in new-generation foods: Challenges, technological effects, biological activity, and regulatory issues. *Crit. Rev. Food Sci. Nutr.* **2022**, *62*, 4929–4950. [CrossRef] [PubMed]
2. Vieira, M.V.; Turkiewicz, I.P.; Tkacz, K.; Fuentes-Grünewald, C.; Pastrana, L.M.; Fuciños, P.; Wojdyło, A.; Nowicka, P. Microalgae as a Potential Functional Ingredient: Evaluation of the Phytochemical Profile, Antioxidant Activity and In-Vitro Enzymatic Inhibitory Effect of Different Species. *Molecules* **2021**, *26*, 7593. [CrossRef] [PubMed]
3. Borowitzka, M.A. Biology of Microalgae. In *Microalgae in Health and Disease Prevention*; Levine, I.A., Fleurence, J., Eds.; Academic Press: Cambridge, MA, USA, 2018; pp. 23–72.
4. Adarme-Vega, T.C.; Thomas-Hall, S.R.; Lim, D.K.Y.; Schenk, P.M. Effects of Long Chain Fatty Acid Synthesis and Associated Gene Expression in Microalga *Tetraselmis* sp. *Mar. Drugs* **2014**, *12*, 3381–3398. [CrossRef]
5. Dahmen-Ben Moussa, I.; Chtourou, H.; Karray, F.; Sayadi, S.; Dhouib, A. Nitrogen or phosphorus repletion strategies for enhancing lipid or carotenoid production from Tetraselmis marina. *Bioresour. Technol.* **2017**, *238*, 325–332. [CrossRef]
6. Cameron, H.; Mata, M.T.; Riquelme, C. The effect of heavy metals on the viability of Tetraselmis marina AC16-MESO and an evaluation of the potential use of this microalga in bioremediation. *Peer J.* **2018**, *6*, 9.e5295. [CrossRef] [PubMed]
7. European Union. Commission implementing regulation (EU) 2017/2470 of 20 December 2017 establishing the Union list of novel foods in accordance with Regulation (EU) 2015/2283 of the European Parliament and of the Council on novel foods. *Off. J. Eur. Union* **2017**, *L351*, 72–201.
8. AECOSAN. Report of the Scientific Committee of the Spanish Agency for Consumer Affairs, Food Safety and Nutrition (AECOSAN) on a request for initial assessment for marketing of the dried marine microalgae Tetraselmis chuii in food supplements under Regulation (EC) No 258/97 on novel foods and novel food ingredients. *AECOSAN Sci. Comm. J.* **2017**, *25*, 11–21.
9. Al-Hoqani, U.; Young, R.; Purton, S. The biotechnological potential of Nannochloropsis. *Perspect. Phycol.* **2017**, *4*, 1–15. [CrossRef]
10. Lubián, L.M.; Montero, O.; Moreno-Garrido, I.; Huertas, I.E.; Sobrino, C.; González-del Valle, M.; Parés, G. Nannochloropsis (Eustigmatophyceae) as source of commercially valuable pigments. *J. Appl. Phycol.* **2000**, *12*, 249–255. [CrossRef]
11. Olofsson, M.; Lamela, T.; Nilsson, E.; Bergé, J.-P.; Del Pino, V.; Uronen, P.; Legrand, C. Combined Effects of Nitrogen Concentration and Seasonal Changes on the Production of Lipids in Nannochloropsis oculata. *Mar. Drugs* **2014**, *12*, 1891–1910. [CrossRef] [PubMed]
12. Van Vooren, G.; Le Grand, F.; Legrand, J.; Cuiné, S.; Peltier, G.; Pruvost, J. Investigation of fatty acids accumulation in Nannochloropsis oculata for biodiesel application. *Bioresour. Technol.* **2012**, *124*, 421–432. [CrossRef] [PubMed]
13. Kim, S.Y.; Kwon, Y.M.; Kim, K.W.; Kim, J.Y.H. Exploring the Potential of Nannochloropsis sp. Extract for Cosmeceutical Applications. *Mar. Drugs* **2021**, *19*, 690. [CrossRef] [PubMed]
14. Kumar, R.; Hegde, A.S.; Sharma, K.; Parmar, P.; Srivatsan, V. Microalgae as a sustainable source of edible proteins and bioactive peptides—Current trends and future prospects. *Food Res. Int.* **2022**, *157*, 111338. [CrossRef] [PubMed]
15. Levasseur, W.; Perré, P.; Pozzobon, V. A review of high value-added molecules production by microalgae in light of the classification. *Biotechnol. Adv.* **2020**, *41*, 107545. [CrossRef]
16. Pereira, H.; Silva, J.; Santos, T.; Gangadhar, K.N.; Raposo, A.; Nunes, C.; Coimbra, M.A.; Gouveia, L.; Barreira, L.; Varela, J. Nutritional Potential and Toxicological Evaluation of Tetraselmis sp. CTP4 Microalgal Biomass Produced in Industrial Photobioreactors. *Molecules* **2019**, *24*, 3192. [CrossRef] [PubMed]

17. Zanella, L.; Vianello, F. Microalgae of the genus Nannochloropsis: Chemical composition and functional implications for human nutrition. *J. Funct. Foods* **2020**, *68*, 103919. [CrossRef]
18. Cavonius, L.R.; Albers, E.; Undeland, I. In vitro bioaccessibility of proteins and lipids of pH-shift processed Nannochloropsis oculata microalga. *Food Funct.* **2016**, *7*, 2016–2024. [CrossRef]
19. Vizcaíno, A.J.; Sáez, M.I.; Martínez, T.F.; Acién, F.G.; Alarcón, F.J. Differential hydrolysis of proteins of four microalgae by the digestive enzymes of gilthead sea bream and Senegalese sole. *Algal Res.* **2019**, *37*, 145–153. [CrossRef]
20. Qazi, W.M.; Ballance, S.; Uhlen, A.K.; Kousoulaki, K.; Haugen, J.-E.; Rieder, A. Protein enrichment of wheat bread with the marine green microalgae Tetraselmis chuii—Impact on dough rheology and bread quality. *LWT* **2021**, *143*, 111115. [CrossRef]
21. Sandgruber, F.; Gielsdorf, A.; Baur, A.C.; Schenz, B.; Müller, S.M.; Schwerdtle, T.; Stangl, G.I.; Griehl, C.; Lorkowski, S.; Dawczynski, C. Variability in Macro- and Micronutrients of 15 Commercially Available Microalgae Powders. *Mar. Drugs* **2021**, *19*, 310. [CrossRef] [PubMed]
22. Wild, K.J.; Steingaß, H.; Rodehutscord, M. Variability in nutrient composition and in vitro crude protein digestibility of 16 microalgae products. *J. Anim. Physiol. Anim. Nutr.* **2018**, *102*, 1306–1319. [CrossRef] [PubMed]
23. Ma, X.N.; Chen, T.P.; Yang, B.; Liu, J.; Chen, F. Lipid Production from Nannochloropsis. *Mar. Drugs* **2016**, *14*, 61. [CrossRef] [PubMed]
24. Moser, G.A.O.; Barrera-Alba, J.J.; Ortega, M.J.; Alves-de-Souza, C.; Bartual, A. Comparative characterization of three Tetraselmis chui (Chlorophyta) strains as sources of nutraceuticals. *J. Appl. Phycol.* **2022**, *34*, 821–835. [CrossRef]
25. Volkman, J.K.; Brown, M.R.; Dunstan, G.A.; Jeffrey, S.W. The biochemical composition of marine microalgae from the class Eustigmatophyceae. *J. Phycol.* **1993**, *29*, 69–78. [CrossRef]
26. Krienitz, L.; Wirth, M. The high content of polyunsaturated fatty acids in Nannochloropsis limnetica (Eustigmatophyceae) and its implication for food web interactions, freshwater aquaculture and biotechnology. *Limnologica* **2006**, *36*, 204–210. [CrossRef]
27. Verspreet, J.; Soetemans, L.; Gargan, C.; Hayes, M.; Bastiaens, L. Nutritional Profiling and Preliminary Bioactivity Screening of Five Micro-Algae Strains Cultivated in Northwest Europe. *Foods* **2021**, *10*, 1516. [CrossRef] [PubMed]
28. Banskota, A.H.; Sperker, S.; Stefanova, R.; McGinn, P.J.; O'Leary, S.J.B. Antioxidant properties and lipid composition of selected microalgae. *J. Appl. Phycol.* **2019**, *31*, 309–318. [CrossRef]
29. Patil, V.; Källqvist, T.; Olsen, E.; Vogt, G.; Gislerød, H.R. Fatty acid composition of 12 microalgae for possible use in aquaculture feed. *Aquac. Int.* **2007**, *15*, 1–9. [CrossRef]
30. Niccolai, A.; Chini Zittelli, G.; Rodolfi, L.; Biondi, N.; Tredici, M.R. Microalgae of interest as food source: Biochemical composition and digestibility. *Algal Res.* **2019**, *42*, 101617. [CrossRef]
31. Bartley, M.L.; Boeing, W.J.; Daniel, D.; Dungan, B.N.; Schaub, T. Optimization of environmental parameters for Nannochloropsis salina growth and lipid content using the response surface method and invading organisms. *J. Appl. Phycol.* **2016**, *28*, 15–24. [CrossRef]
32. Rebolloso-Fuentes, M.M.; Navarro-Pérez, A.; García-Camacho, F.; Ramos-Miras, J.J.; Guil-Guerrero, J.L. Biomass nutrient profiles of the microalga Nannochloropsis. *J. Agric. Food Chem.* **2001**, *49*, 2966–2972. [CrossRef]
33. Van Wagenen, J.; Miller, T.W.; Hobbs, S.; Hook, P.; Crowe, B.; Huesemann, M. Effects of Light and Temperature on Fatty Acid Production in Nannochloropsis Salina. *Energies* **2012**, *5*, 731–740. [CrossRef]
34. Ahmmed, M.K.; Ahmmed, F.; Tian, H.; Carne, A.; Bekhit, A.E.-D. Marine omega-3 (n-3) phospholipids: A comprehensive review of their properties, sources, bioavailability, and relation to brain health. *Compr. Rev. Food Sci. Food Saf.* **2020**, *19*, 64–123. [CrossRef] [PubMed]
35. Couto, D.; Conde, T.A.; Melo, T.; Neves, B.; Costa, M.; Cunha, P.; Guerra, I.; Correia, N.; Silva, J.T.; Pereira, H.; et al. Effects of outdoor and indoor cultivation on the polar lipid composition and antioxidant activity of Nannochloropsis oceanica and Nannochloropsis limnetica: A lipidomics perspective. *Algal Res.* **2022**, *64*, 102718. [CrossRef]
36. Mitra, M.; Patidar, S.K.; Mishra, S. Integrated process of two stage cultivation of Nannochloropsis sp. for nutraceutically valuable eicosapentaenoic acid along with biodiesel. *Bioresour. Technol.* **2015**, *193*, 363–369. [CrossRef] [PubMed]
37. Grubišic, M.; Šantek, B.; Zoric, Z.; Cošic, Z.; Vrana, I.; Gašparovic, B.; Což-Rakovac, R.; Šantek, M.I. Bioprospecting of Microalgae Isolated from the Adriatic Sea: Characterization of Biomass, Pigment, Lipid and Fatty Acid Composition, and Antioxidant and Antimicrobial Activity. *Molecules* **2022**, *27*, 1248. [CrossRef]
38. Banskota, A.H.; Stefanova, R.; Sperker, S.; McGinn, P.J. New diacylglyceryltrimethylhomoserines from the marine microalga Nannochloropsis granulata and their nitric oxide inhibitory activity. *J. Appl. Phycol.* **2013**, *25*, 1513–1521. [CrossRef]
39. Kagan, M.L.; West, A.L.; Zante, C.; Calder, P.C. Acute appearance of fatty acids in human plasma—A comparative study between polar-lipid rich oil from the microalgae Nannochloropsis oculata and krill oil in healthy young males. *Lipids Health Dis.* **2013**, *12*, 102. [CrossRef]
40. Rao, A.; Briskey, D.; Nalley, J.O.; Ganuza, E. Omega-3 Eicosapentaenoic Acid (EPA) Rich Extract from the Microalga Nannochloropsis Decreases Cholesterol in Healthy Individuals: A Double-Blind, Randomized, Placebo-Controlled, Three-Month Supplementation Study. *Nutrients* **2020**, *12*, 1869. [CrossRef] [PubMed]
41. Brown, M.R. The amino-acid and sugar composition of 16 species of microalgae used in mariculture. *J. Exp. Mar. Biol. Ecol.* **1991**, *145*, 79–99. [CrossRef]

42. Di Lena, G.; Casini, I.; Lucarini, M.; Sanchez del Pulgar, J.; Aguzzi, A.; Caproni, R.; Gabrielli, P.; Lombardi-Boccia, G. Chemical characterization and nutritional evaluation of microalgal biomass from large-scale production: A comparative study of five species. *Eur. Food Res. Technol.* **2020**, *246*, 323–332. [CrossRef]
43. Vogler, B.W.; Brannum, J.; Chung, J.W.; Seger, M.; Posewitz, M.C. Characterization of the Nannochloropsis gaditana storage carbohydrate: A 1,3-beta glucan with limited 1,6-branching. *Algal Res.* **2018**, *36*, 152–158. [CrossRef]
44. Espinoza, D.; Contreras-Porcia, L.; Ehrenfeld, N. B-glucans, production and properties in microalgae with emphasis on Nannochloropsis genus (Ochrophyta, Eustigmatales). *Rev. Biol. Mar. Oceanogr.* **2017**, *52*, 33–49.
45. Becker, B.; Becker, D.; Kamerling, J.P.; Melkonian, M. 2-Keto-sugar acids in green flagellates: A chemical marker for Prasinophycean scales. *J. Phycol.* **1991**, *27*, 498–504. [CrossRef]
46. Kermanshahi-Pour, A.; Sommer, T.J.; Anastas, P.T.; Zimmerman, J.B. Enzymatic and acid hydrolysis of Tetraselmis suecica for polysaccharide characterization. *Bioresour. Technol.* **2014**, *173*, 415–421. [CrossRef] [PubMed]
47. Templeton, D.W.; Quinn, M.; Van Wychen, S.; Hyman, D.; Laurens, L.M.L. Separation and quantification of microalgal carbohydrates. *J. Chromatogr. A* **2012**, *1270*, 225–234. [CrossRef]
48. Tibbetts, S.M.; Milley, J.E.; Lall, S.P. Chemical composition and nutritional properties of freshwater and marine microalgal biomass cultured in photobioreactors. *J. Appl. Phycol.* **2015**, *27*, 1109–1119. [CrossRef]
49. Ampofo, J.; Abbey, L. Microalgae: Bioactive Composition, Health Benefits, Safety and Prospects as Potential High-Value Ingredients for the Functional Food Industry. *Foods* **2022**, *11*, 1744. [CrossRef] [PubMed]
50. Del Mondo, A.; Smerilli, A.; Sane, E.; Sansone, C.; Brunet, C. Challenging microalgal vitamins for human health. *Microb. Cell Factories* **2020**, *19*, 201. [CrossRef]
51. De Roeck-Holtzhauer, Y.; Quere, I.; Claire, C. Vitamin analysis of five planktonic microalgae and one macroalga. *J. Appl. Phycol.* **1991**, *3*, 259–264. [CrossRef]
52. Fabregas, J.; Herrero, C. Vitamin content of four marine microalgae. Potential use as source of vitamins in nutrition. *J. Ind. Microbiol.* **1990**, *5*, 259–264. [CrossRef]
53. Santiago-Morales, I.S.; Trujillo-Valle, L.; Márquez-Rocha, F.J.; López Hernández, J.F. Tocopherols, Phycocyanin and Superoxide Dismutase from Microalgae: As Potential Food Antioxidants. *Appl. Food Biotechnol.* **2018**, *5*, 19–27. [CrossRef]
54. Brown, M.R.; Mular, M.; Miller, I.; Farmer, C.; Trenerry, C. The vitamin content of microalgae used in aquaculture. *J. Appl. Phycol.* **1999**, *11*, 247–255. [CrossRef]
55. Edelmann, M.; Aalto, S.; Chamlagain, B.; Kariluoto, S.; Piironen, V. Riboflavin, niacin, folate and vitamin B12 in commercial microalgae powders. *J. Food Comp. Anal.* **2019**, *82*, 103226. [CrossRef]
56. Durmaz, Y. Vitamin E (α-tocopherol) production by the marine microalgae Nannochloropsis oculata (Eustigmatophyceae) in nitrogen limitation. *Aquaculture* **2007**, *272*, 717–722. [CrossRef]
57. Safafar, H.; van Wagenen, J.; Møller, P.; Jacobsen, C. Carotenoids, Phenolic Compounds and Tocopherols Contribute to the Antioxidative Properties of Some Microalgae Species Grown on Industrial Wastewater. *Mar. Drugs* **2015**, *13*, 7339–7356. [CrossRef]
58. Ljubic, A.; Jacobsen, C.; Holdt, S.L.; Jakobsen, J. Microalgae Nannochloropsis oceanica as a future new natural source of vitamin D3. *Food Chem.* **2020**, *320*, 126627. [CrossRef]
59. Goiris, K.; Van Colen, W.; Wilches, I.; Leon-Tamariz, F.; De Cooman, L.; Muylaert, K. Impact of nutrient stress on antioxidant production in three species of microalgae. *Algal Res.* **2015**, *7*, 51–57. [CrossRef]
60. Di Lena, G.; Casini, I.; Lucarini, M.; Lombardi-Boccia, G. Carotenoid profiling of five microalgae species from large-scale production. *Food Res. Int.* **2019**, *120*, 810–818. [CrossRef]
61. Sathasivam, R.; Ki, J.S. A Review of the Biological Activities of Microalgal Carotenoids and Their Potential Use in Healthcare and Cosmetic Industries. *Mar. Drugs* **2018**, *16*, 26. [CrossRef]
62. Ahmed, F.; Fanning, K.; Netzel, M.; Turner, W.; Li, Y.; Schenk, P.M. Profiling of carotenoids and antioxidant capacity of microalgae from subtropical coastal and brackish waters. *Food Chem.* **2014**, *165*, 300–306. [CrossRef] [PubMed]
63. Ryckebosch, E.; Bruneel, C.; Termote-Verhalle, R.; Muylaert, K.; Foubert, I. Influence of extraction solvent system on extractability of lipid components from different microalgae species. *Algal Res.* **2014**, *3*, 36–43. [CrossRef]
64. Maadane, A.; Merghoub, N.; Ainane, T.; El Arroussi, H.; Benhima, R.; Amzazi, S.; Bakri, Y.; Wahby, I. Antioxidant activity of some Moroccan marine microalgae: Pufa profiles, carotenoids and phenolic content. *J. Biotechnol.* **2015**, *215*, 13–19. [CrossRef]
65. Goiris, K.; Muylaert, K.; Fraeye, I.; Foubert, I.; De Brabanter, J.; De Cooman, L. Antioxidant potential of microalgae in relation to their phenolic and carotenoid content. *J. Appl. Phycol.* **2012**, *24*, 1477–1486. [CrossRef]
66. Cardoso, C.; Pereira, H.; Franca, J.; Matos, J.; Monteiro, I.; Pousao-Ferreira, P.; Gomes, A.; Barreira, L.; Varela, J.; Neng, N.; et al. Lipid composition and some bioactivities of 3 newly isolated microalgae (Tetraselmis sp. IMP3, Tetraselmis sp. CTP4, and Skeletonema sp.). *Aquac. Int.* **2020**, *28*, 711–727. [CrossRef]
67. Scaglioni, P.T.; Quadros, L.; de Paula, M.; Furlong, V.B.; Abreu, P.C.; Badiale-Furlong, E. Inhibition of Enzymatic and Oxidative Processes by Phenolic Extracts from Spirulina sp. and Nannochloropsis sp. *Food Technol. Biotechnol.* **2018**, *56*, 344–353. [CrossRef] [PubMed]
68. Haoujar, I.; Cacciola, F.; Abrini, J.; Mangraviti, D.; Giuffrida, D.; El Majdoub, Y.O.; Kounnoun, A.; Miceli, N.; Taviano, M.F.; Mondello, L.; et al. The Contribution of Carotenoids, Phenolic Compounds, and Flavonoids to the Antioxidative Properties of Marine Microalgae Isolated from Mediterranean Morocco. *Molecules* **2019**, *24*, 4037. [CrossRef] [PubMed]

69. Custódio, L.; Justo, T.; Silvestre, L.; Barradas, A.; Duarte, C.V.; Pereira, H.; Barreira, L.; Rauter, A.P.; Albericio, F.; Varela, J. Microalgae of different phyla display antioxidant, metal chelating and acetylcholinesterase inhibitory activities. *Food Chem.* **2012**, *131*, 134–140. [CrossRef]
70. Dolganyuk, V.; Belova, D.; Babich, O.; Prosekov, A.; Ivanova, S.; Katserov, D.; Patyukov, N.; Sukhikh, S. Microalgae: A Promising Source of Valuable Bioproducts. *Biomolecules* **2020**, *10*, 1153. [CrossRef]
71. Barkia, I.; Saari, N.; Manning, S.R. Microalgae for High-Value Products Towards Human Health and Nutrition. *Mar. Drugs* **2019**, *17*, 304. [CrossRef] [PubMed]
72. Listing the World's Algae. Available online: www.algaebase.org (accessed on 18 October 2022).
73. Pham-Huy, L.A.; He, H.; Pham-Huy, C. Free radicals, antioxidants in disease and health. *Int. J. Biomed. Sci.* **2008**, *4*, 89–96.
74. Sen, S.; Chakraborty, R.; Sridhar, C.; Reddy, Y.; De, B. Free radicals, antioxidants, diseases and phytomedicines: Current status and future prospect. *Int. J. Pharm. Sci. Rev. Res.* **2010**, *3*, 91–100.
75. Pashkow, F.J. Oxidative Stress and Inflammation in Heart Disease: Do Antioxidants Have a Role in Treatment and/or Prevention? *Int. J. Inflamm.* **2011**, *2011*, 514623. [CrossRef]
76. Conde, T.A.; Neves, B.F.; Couto, D.; Melo, T.; Neves, B.; Costa, M.; Silva, J.; Domingues, P.; Domingues, M.R. Microalgae as Sustainable Bio-Factories of Healthy Lipids: Evaluating Fatty Acid Content and Antioxidant Activity. *Mar. Drugs* **2021**, *19*, 357. [CrossRef] [PubMed]
77. Coulombier, N.; Jauffrais, T.; Lebouvier, N. Antioxidant Compounds from Microalgae: A Review. *Mar. Drugs* **2021**, *19*, 549. [CrossRef] [PubMed]
78. Leopoldini, M.; Russo, N.; Toscano, M. The molecular basis of working mechanism of natural polyphenolic antioxidants. *Food Chem.* **2011**, *125*, 288–306. [CrossRef]
79. Custódio, L.; Soares, F.; Pereira, H.; Barreira, L.; Vizetto-Duarte, C.; Rodrigues, M.J.; Rauter, A.P.; Alberício, F.; Varela, J. Fatty acid composition and biological activities of Isochrysis galbana T-ISO, Tetraselmis sp. and Scenedesmus sp.: Possible application in the pharmaceutical and functional food industries. *J. Appl. Phycol.* **2014**, *26*, 151–161. [CrossRef]
80. Coulombier, N.; Nicolau, E.; Le Déan, L.; Antheaume, C.; Jauffrais, T.; Lebouvier, N. Impact of Light Intensity on Antioxidant Activity of Tropical Microalgae. *Mar. Drugs* **2020**, *18*, 122. [CrossRef]
81. Amna Kashif, S.; Hwang, Y.J.; Park, J.K. Potent biomedical applications of isolated polysaccharides from marine microalgae Tetraselmis species. *Bioprocess Biosyst. Eng.* **2018**, *41*, 1611–1620. [CrossRef] [PubMed]
82. Gam, D.H.; Yi Kim, S.; Kim, J.W. Optimization of Ultrasound-Assisted Extraction Condition for Phenolic Compounds, Antioxidant Activity, and Epigallocatechin Gallate in Lipid-Extracted Microalgae. *Molecules* **2020**, *25*, 454. [CrossRef] [PubMed]
83. Kokkali, M.; Martí-Quijal, F.J.; Taroncher, M.; Ruiz, M.J.; Kousoulaki, K.; Barba, F.J. Improved Extraction Efficiency of Antioxidant Bioactive Compounds from Tetraselmis chuii and Phaedoactylum tricornutum Using Pulsed Electric Fields. *Molecules* **2020**, *25*, 3921. [CrossRef]
84. Silva, M.; Kamberovic, F.; Uota, S.T.; Kovan, I.-M.; Viegas, C.S.B.; Simes, D.C.; Gangadhar, K.N.; Varela, J.; Barreira, L. Microalgae as Potential Sources of Bioactive Compounds for Functional Foods and Pharmaceuticals. *Appl. Sci.* **2022**, *12*, 5877. [CrossRef]
85. Trentin, R.; Custódio, L.; Rodrigues, M.J.; Moschin, E.; Sciuto, K.; da Silva, J.P.; Moro, I. Total Phenolic Levels, In Vitro Antioxidant Properties, and Fatty Acid Profile of Two Microalgae, Tetraselmis marina Strain IMA043 and Naviculoid Diatom Strain IMA053, Isolated from the North Adriatic Sea. *Mar. Drugs* **2022**, *20*, 207. [CrossRef]
86. Lee, K.H.; Jang, Y.W.; Kim, H.; Ki, J.-S.; Yoo, H.Y. Optimization of Lutein Recovery from Tetraselmis suecica by Response Surface Methodology. *Biomolecules* **2021**, *11*, 182. [CrossRef] [PubMed]
87. Rentería-Mexía, A.; Ulloa-Mercado, G.; Gortáres-Moroyoqui, P.; González-Mercado, A.; Sánchez, M.; Sineiro, J.; Núñez, M.J. Antioxidant potential and antiangiogenic activity of Tetraselmis suecica grown in a semicontinuous culture. *J. Chem. Technol. Biotechnol.* **2022**, *97*, 2528–2536. [CrossRef]
88. Sansone, C.; Galasso, C.; Orefice, I.; Nuzzo, G.; Luongo, E.; Cutignano, A.; Romano, G.; Brunet, C.; Fontana, A.; Esposito, F.; et al. The green microalga Tetraselmis suecica reduces oxidative stress and induces repairing mechanisms in human cells. *Sci. Rep.* **2017**, *7*, 41215. [CrossRef] [PubMed]
89. Lauritano, C.; Andersen, J.H.; Hansen, E.; Albrigtsen, M.; Escalera, L.; Esposito, F.; Helland, K.; Hanssen, K.Ø.; Romano, G.; Ianora, A. Bioactivity Screening of Microalgae for Antioxidant, Anti-Inflammatory, Anticancer, Anti-Diabetes, and Antibacterial Activities. *Front. Mar. Sci.* **2016**, *3*, 68. [CrossRef]
90. Norzagaray-Valenzuela, C.D.; Valdez-Ortiz, A.; Shelton, L.M.; Jiménez-Edeza, M.; Rivera-López, J.; Valdez-Flores, M.A.; Germán-Báez, L.J. Residual biomasses and protein hydrolysates of three green microalgae species exhibit antioxidant and anti-aging activity. *J. Appl. Phycol.* **2017**, *29*, 189–198. [CrossRef]
91. Parra-Riofrío, G.; García-Márquez, J.; Casas-Arrojo, V.; Uribe-Tapia, E.; Abdala-Díaz, R.T. Antioxidant and Cytotoxic Effects on Tumor Cells of Exopolysaccharides from Tetraselmis suecica (Kylin) Butcher Grown Under Autotrophic and Heterotrophic Conditions. *Mar. Drugs* **2020**, *18*, 534. [CrossRef] [PubMed]
92. Natrah, F.M.I.; Yusoff, F.M.; Shariff, M.; Abas, F.; Mariana, N.S. Screening of Malaysian indigenous microalgae for antioxidant properties and nutritional value. *J. Appl. Phycol.* **2007**, *19*, 711–718. [CrossRef]
93. Deepa, P.K.; Subramanian, A.; Manjusha, W.A. Evaluation of antioxidant potential and bioactive metabolites of Nannochloropsis sp. *J. Adv. Appl. Sci. Res.* **2021**, *2*, 1–9. [CrossRef]

94. Khatib, S.; Artoul, F.; Paluy, I.; Boluchevsky, L.; Kvitnitsky, E.; Vaya, J. Nannochloropsis sp. ethanol extract prevents macrophage and LDL oxidation and enhances PON1 activity through the principal active compound lyso-diacylglyceryltrimethylhomoserine (lyso-DGTS). *J. Appl. Phycol.* **2018**, *30*, 1679–1689. [CrossRef]
95. Santhar, D.T.; Haq, M.A.B.; Marudhupandi, T.; Vaseeharan, B.; Rajan, D.K.; Moovendhan, M. Evaluation of chemical compositions and antioxidant potential of marine microalgae of the genus Nannochloropsis. *Biomass Convers. Biorefinery* **2021**. [CrossRef]
96. Goh, S.H.; Yusoff, F.M.; Loh, S.P. A comparison of the antioxidant properties and total phenolic content in a Diatom, Chaetoceros sp. and a green microalga, Nannochloropsis sp. *J. Agric. Sci.* **2010**, *2*, 123–130. [CrossRef]
97. Guedes, A.C.; Gião, M.S.; Seabra, R.; Ferreira, A.C.S.; Tamagnini, P.; Moradas-Ferreira, P.; Malcata, F.X. Evaluation of the Antioxidant Activity of Cell Extracts from Microalgae. *Mar. Drugs* **2013**, *11*, 1256–1270. [CrossRef] [PubMed]
98. Pereira, H.; Custódio, L.; Rodrigues, M.J.; De Sousa, C.B.; Oliveira, M.; Barreira, L.; Neng, N.D.R.; Nogueira, J.M.F.; Alrokayan, S.A.; Mouffouk, F.; et al. Biological Activities and Chemical Composition of Methanolic Extracts of Selected Autochthonous Microalgae Strains from the Red Sea. *Mar. Drugs* **2015**, *13*, 3531–3549. [CrossRef] [PubMed]
99. Martínez, R.; García-Beltrán, A.; Kapravelou, G.; Mesas, C.; Cabeza, L.; Perazzoli, G.; Guarnizo, P.; Rodríguez-López, A.; Andrés Vallejo, R.; Galisteo, M.; et al. In Vivo Nutritional Assessment of the Microalga Nannochloropsis gaditana and Evaluation of the Antioxidant and Antiproliferative Capacity of Its Functional Extracts. *Mar. Drugs* **2022**, *20*, 318. [CrossRef] [PubMed]
100. Medina, C.; Rubilar, M.; Shene, C.; Torres, S.; Verdugo, M. Protein Fractions with Techno-Functional and Antioxidant Properties from Nannochloropsis gaditana Microalgal Biomass. *J. Biobased Mater. Bioenergy* **2015**, *9*, 417–425. [CrossRef]
101. Millao, S.; Uquiche, E. Antioxidant activity of supercritical extracts from Nannochloropsis gaditana: Correlation with its content of carotenoids and tocopherols. *J. Supercrit. Fluids* **2016**, *111*, 143–150. [CrossRef]
102. Ebrahimzadeh, M.A.; Khalili, M.; Dehpour, A.A. Antioxidant activity of ethyl acetate and methanolic extracts of two marine algae, *Nannochloropsis oculata* and *Gracilaria gracilis*—An in vitro assay. *Braz. J. Pharm. Sci.* **2018**, *54*, e17280. [CrossRef]
103. Feller, R.; Matos, Â.P.; Mazzutti, S.; Moecke, E.H.S.; Tres, M.V.; Derner, R.B.; Oliveira, J.V.; Junior, A.F. Polyunsaturated ω-3 and ω-6 fatty acids, total carotenoids and antioxidant activity of three marine microalgae extracts obtained by supercritical CO2 and subcritical n-butane. *J. Supercrit. Fluids* **2018**, *133*, 437–443. [CrossRef]
104. Ben Hafsa, M.; Ben Ismail, M.; Garrab, M.; Aly, R.; Gagnon, J.; Naghmouchi, K. Antimicrobial, antioxidant, cytotoxic and anticholinesterase activities of water-soluble polysaccharides extracted from microalgae Isochrysis galbana and Nannochloropsis oculata. *J. Serbian Chem. Soc.* **2017**, *82*, 509–522. [CrossRef]
105. Hussein, H.A.; Mohamad, H.; Mohd Ghazaly, M.; Laith, A.A.; Abdullah, M.A. Anticancer and antioxidant activities of Nannochloropsis oculata and Chlorella sp. extracts in co-application with silver nanoparticle. *J. King Saud Univ. Sci.* **2020**, *32*, 3486–3494. [CrossRef]
106. Wali, A.F.; Al Dhaheri, Y.; Ramakrishna Pillai, J.; Mushtaq, A.; Rao, P.G.M.; Rabbani, S.A.; Firdous, A.; Elshikh, M.S.; Farraj, D.A.A. LC-MS Phytochemical Screening, In Vitro Antioxidant, Antimicrobial and Anticancer Activity of Microalgae Nannochloropsis oculata Extract. *Separations* **2020**, *7*, 54. [CrossRef]
107. Stahl, W.; Sies, H. Photoprotection by dietary carotenoids: Concept, mechanisms, evidence and future development. *Mol. Nutr. Food Res.* **2012**, *56*, 287–295. [CrossRef]
108. El-Agamey, A.; McGarvey, D.J. Carotenoid Radicals and Radical Ions. In *Carotenoids: Volume 4: Natural Functions*; Britton, G., Liaaen-Jensen, S., Pfander, H., Eds.; Birkhäuser Basel: Basel, Switzerland, 2008; pp. 119–154.
109. Ochoa Becerra, M.; Mojica Contreras, L.; Hsieh Lo, M.; Mateos Díaz, J.; Castillo Herrera, G. Lutein as a functional food ingredient: Stability and bioavailability. *J. Funct. Foods* **2020**, *66*, 103771. [CrossRef]
110. de Jesus Raposo, M.F.; de Morais, A.M.M.B.; de Morais, R.M.S.C. Bioactivity and Applications of Polysaccharides from Marine Microalgae. In *Polysaccharides: Bioactivity and Biotechnology*; Ramawat, K.G., Mérillon, J.-M., Eds.; Springer International Publishing: Cham, Switzerland, 2021; pp. 1–38.
111. Qiao, H.; Hu, D.; Ma, J.; Wang, X.; Wu, H.; Wang, J. Feeding effects of the microalga Nannochloropsis sp. on juvenile turbot (*Scophthalmus maximus* L.). *Algal Res.* **2019**, *41*, 101540. [CrossRef]
112. Rahman, N.A.; Khatoon, H.; Yusuf, N.; Banerjee, S.; Haris, N.A.; Lananan, F.; Tomoyo, K. Tetraselmis chuii biomass as a potential feed additive to improve survival and oxidative stress status of Pacific white-leg shrimp Litopenaeus vannamei postlarvae. *Int. Aquat. Res.* **2017**, *9*, 235–247. [CrossRef]
113. Nacer, W.; Baba Ahmed, F.Z.; Merzouk, H.; Benyagoub, O.; Bouanane, S. Evaluation of the anti-inflammatory and antioxidant effects of the microalgae Nannochloropsis gaditana in streptozotocin-induced diabetic rats. *J. Diabetes Metab. Disord.* **2020**, *19*, 1483–1490. [CrossRef]
114. Abu-Ghannam, N.; Rajauria, G. Antimicrobial activity of compounds isolated from algae. In *Functional Ingredients from Algae for Foods and Nutraceuticals*; Domínguez, H., Ed.; Woodhead Publishing: Sawston, UK, 2013; pp. 287–306.
115. Herrero, M.; Ibáñez, E.; Cifuentes, A.; Reglero, G.; Santoyo, S. Dunaliella salina Microalga Pressurized Liquid Extracts as Potential Antimicrobials. *J. Food Prot.* **2006**, *69*, 2471–2477. [CrossRef]
116. Klejdus, B.; Lojková, L.; Plaza, M.; Šnóblová, M.; Štěrbová, D. Hyphenated technique for the extraction and determination of isoflavones in algae: Ultrasound-assisted supercritical fluid extraction followed by fast chromatography with tandem mass spectrometry. *J. Chromat. A* **2010**, *1217*, 7956–7965. [CrossRef]
117. Bhagavathy, S.; Sumathi, P.; Jancy Sherene Bell, I. Green algae Chlorococcum humicola-a new source of bioactive compounds with antimicrobial activity. *Asian Pac. J. Trop. Biomed.* **2011**, *1*, S1–S7. [CrossRef]

118. Maadane, A.; Merghoub, N.; El Mernissi, N.; Ainane, T.; Amzazi, S.; Wahby, I.; Bakri, Y. Antimicrobial activity of marine microalgae isolated from Moroccan coastlines. *J. Microbiol. Biotechnol. Food Sci.* **2017**, *6*, 1257–1260. [CrossRef]
119. Jusidin, M.R.; Othman, R.; Shaleh, S.R.M.; Ching, F.F.; Senoo, S.; Oslan, S.N.H. In Vitro Antibacterial Activity of Marine Microalgae Extract against *Vibrio harveyi*. *Appl. Sci.* **2022**, *12*, 1148. [CrossRef]
120. Austin, B.; Baudet, E.; Stobie, M. Inhibition of bacterial fish pathogens by Tetraselmis suecica. *J. Fish Dis.* **1992**, *15*, 55–61. [CrossRef]
121. Khavari, F.; Saidijam, M.; Taheri, M.; Nouri, F. Microalgae: Therapeutic potentials and applications. *Mol. Biol. Rep.* **2021**, *48*, 4757–4765. [CrossRef]
122. Guzman, F.; Wong, G.; Roman, T.; Cardenas, C.; Alvarez, C.; Schmitt, P.; Albericio, F.; Rojas, V. Identification of Antimicrobial Peptides from the Microalgae Tetraselmis suecica (Kylin) Butcher and Bactericidal Activity Improvement. *Mar. Drugs* **2019**, *17*, 453. [CrossRef]
123. Hoernke, M.; Schwieger, C.; Kerth, A.; Blume, A. Binding of cationic pentapeptides with modified side chain lengths to negatively charged lipid membranes: Complex interplay of electrostatic and hydrophobic interactions. *Biochim. Biophys. Acta Biomembr. BBA* **2012**, *1818*, 1663–1672. [CrossRef] [PubMed]
124. He, F.; Yang, Y.; Yang, G.; Yu, L. Studies on antibacterial activity and antibacterial mechanism of a novel polysaccharide from Streptomyces virginia H03. *Food Control* **2010**, *21*, 1257–1262. [CrossRef]
125. Ratovitski, E.A. Anticancer Natural Compounds as Epigenetic Modulators of Gene Expression. *Curr. Genom.* **2017**, *18*, 175–205. [CrossRef]
126. Hussein, H.A.; Mohamad, H.; Ghazaly, M.M.; Laith, A.A.; Abdullah, M.A. Cytotoxic effects of Tetraselmis suecica chloroform extracts with silver nanoparticle co-application on MCF-7, 4 T1, and Vero cell lines. *J. Appl. Phycol.* **2020**, *32*, 127–143. [CrossRef]
127. Hussein, H.A.; Kassim, M.N.I.; Maulidiani, M.; Abas, F.; Abdullah, M.A. Cytotoxicity and 1H NMR metabolomics analyses of microalgal extracts for synergistic application with Tamoxifen on breast cancer cells with reduced toxicity against Vero cells. *Heliyon* **2022**, *8*, e09192. [CrossRef] [PubMed]
128. Sanjeewa, K.K.A.; Fernando, I.P.S.; Samarakoon, K.W.; Lakmal, H.H.C.; Kim, E.-A.; Kwon, O.N.; Dilshara, M.G.; Lee, J.-B.; Jeon, Y.-J. Anti-inflammatory and anti-cancer activities of sterol rich fraction of cultured marine microalga Nannochloropsis oculata. *Algae* **2016**, *31*, 277–287. [CrossRef]
129. Castejón, N.; Marko, D. Fatty Acid Composition and Cytotoxic Activity of Lipid Extracts from Nannochloropsis gaditana Produced by Green Technologies. *Molecules* **2022**, *27*, 3710. [CrossRef] [PubMed]
130. Carrasco-Reinado, R.; Escobar-Niño, A.; Fajardo, C.; Morano, I.M.; Amil-Ruiz, F.; Martinez-Rodríguez, G.; Fuentes-Almagro, C.; Capilla, V.; Tomás-Cobos, L.; Soriano-Romaní, L.; et al. Development of New Antiproliferative Compound against Human Tumor Cells from the Marine Microalgae Nannochloropsis gaditana by Applied Proteomics. *Int. J. Mol. Sci.* **2021**, *22*, 96. [CrossRef] [PubMed]
131. Letsiou, S.; Kalliampakou, K.; Gardikis, K.; Mantecon, L.; Infante, C.; Chatzikonstantinou, M.; Labrou, N.E.; Flemetakis, E. Skin protective effects of Nannochloropsis gaditana extract on H2O2-stressed human dermal fibroblasts. *Front. Mar. Sci.* **2017**, *4*, 221. [CrossRef]
132. Bernaerts, T.M.M.; Verstreken, H.; Dejonghe, C.; Gheysen, L.; Foubert, I.; Grauwet, T.; Van Loey, A.M. Cell disruption of Nannochloropsis sp. improves in vitro bioaccessibility of carotenoids and ω3-LC-PUFA. *J. Funct. Foods* **2020**, *65*, 103770. [CrossRef]
133. EFSA Panel on Biological Hazards. Update of the list of QPS-recommended biological agents intentionally added to food or feed as notified to EFSA 11: Suitability of taxonomic units notified to EFSA until September 2019. *EFSA J.* **2020**, *18*, 5965. [CrossRef]
134. Bernaerts, T.M.M.; Gheysen, L.; Kyomugasho, C.; Jamsazzadeh Kermani, Z.; Vandionant, S.; Foubert, I.; Hendrickx, M.E.; Van Loey, A.M. Comparison of microalgal biomasses as functional food ingredients: Focus on the composition of cell wall related polysaccharides. *Algal Res.* **2018**, *32*, 150–161. [CrossRef]
135. Mantecón, L.; Moyano, R.; Cameán, A.M.; Jos, A. Safety assessment of a lyophilized biomass of Tetraselmis chuii (TetraSOD®) in a 90 day feeding study. *Food Chem. Toxicol.* **2019**, *133*, 110810. [CrossRef]
136. Nunes, M.C.; Fernandes, I.; Vasco, I.; Sousa, I.; Raymundo, A. Tetraselmis chuii as a Sustainable and Healthy Ingredient to Produce Gluten-Free Bread: Impact on Structure, Colour and Bioactivity. *Foods* **2020**, *9*, 579. [CrossRef] [PubMed]
137. Qazi, W.M.; Ballance, S.; Kousoulaki, K.; Uhlen, A.K.; Kleinegris, D.M.M.; Skjånes, K.; Rieder, A. Protein Enrichment of Wheat Bread with Microalgae: Microchloropsis gaditana, Tetraselmis chui and Chlorella vulgaris. *Foods* **2021**, *10*, 3078. [CrossRef] [PubMed]
138. Batista, A.P.; Niccolai, A.; Bursic, I.; Sousa, I.; Raymundo, A.; Rodolfi, L.; Biondi, N.; Tredici, M.R. Microalgae as Functional Ingredients in Savory Food Products: Application to Wheat Crackers. *Foods* **2019**, *8*, 611. [CrossRef] [PubMed]
139. Batista, A.P.; Niccolai, A.; Fradinho, P.; Fragoso, S.; Bursic, I.; Rodolfi, L.; Biondi, N.; Tredici, M.R.; Sousa, I.; Raymundo, A. Microalgae biomass as an alternative ingredient in cookies: Sensory, physical and chemical properties, antioxidant activity and in vitro digestibility. *Algal Res.* **2017**, *26*, 161–171. [CrossRef]

nutrients

MDPI

Article

Impact of the Structural Modifications of Potato Protein in the Digestibility Process under Semi-Dynamic Simulated Human Gastrointestinal In Vitro System

Luis Jiménez-Munoz [1,*], Emmanouil D. Tsochatzis [1,2] and Milena Corredig [1]

[1] Department of Food Science, CiFOOD Center for Innovative Foods, Aarhus University, Agro Food Park 48, 8200 Aarhus, Denmark; emmanouil.tsochatzis@efsa.europa.eu (E.D.T.); mc@food.au.dk (M.C.)
[2] European Food Safety Authority-EFSA, Via Carlo Magno 1A, 43146 Parma, Italy
* Correspondence: author: luisjimenez@food.au.dk

Abstract: The raising consumer demand for plant-derived proteins has led to an increased production of alternative protein ingredients with varying processing histories. In this study, we used a commercially available potato protein ingredient with a nutritionally valuable amino acid profile and high technological functionality to evaluate if the digestibility of a suspension with the same composition is affected by differences in the structure. Four isocaloric (4% protein, *w*/*w*) matrices (suspension, gel, foam and heat-set foam) were prepared and their gastrointestinal fate was followed utilizing a semi-dynamic in vitro digestion model. The microstructure was observed by confocal laser scanning microscopy, protein breakdown was tested by electrophoresis and free amino acids after intestinal digestion was estimated using liquid chromatography/triple-quadruple-mass spectrometry (LC-TQMS). The heat-treated samples showed a higher degree of hydrolysis and lower trypsin inhibitory activity than the non-heat-treated samples. An in vitro digestible indispensable amino acid score was calculated based on experimental data, showing a value of 0.9 based on sulfur amino acids/valine as the limiting amino acids. The heated samples also showed a slower gastric emptying rate. The study highlights the effect of the food matrix on the distribution of the peptides created during various stages of gastric emptying.

Keywords: potato protein isolate; patatin; semi-dynamic in vitro digestion; structuring; amino acid digestibility

1. Introduction

Plant proteins are currently of great interest as they have shown great potential to be used as ingredients in more sustainable and climate friendly foods compared to their traditional animal-derived counterparts [1]. However, more research is needed to understand how to design and offer novel food matrices containing plant proteins able to respond to consumers' demands for wholesome and nutritionally functional foods. Necessarily, a great part of global research so far has focused on the study of extraction and concentration processes aiming to increase the purity of ingredients, to decrease antinutritional components, off-flavor precursors, to improve their technological performance and their nutritional profile. Isolates (≥80% protein) can be obtained with various technologies [2] and each process will result in different effects on the purity, functionality, presence and activity of anti-nutritional factors. Furthermore, protein ingredients, once processed in a food, may have different properties which will affect their digestibility, peptide distributions during gastrointestinal transit, and nutritional bioefficacy [3–5].

Among surging plant-based alternatives, potato protein isolates stand out, as they have the potential to be used as a source of high quality protein for human consumption. Furthermore, their recovery from one of the side-streams of the production of potato starch creates an opportunity to valorize and improve the circularity of this food chain [6].

There are four major classes which compose potato protein extract. Patatin (40 kDa), a glycoprotein (up to 40%), and protease inhibitors (22, 15, 9 kDa) represent up to ~90% [7] of the total extract, with lipoxygenase (130 kDa) making up the rest. The complex trypsin inhibitor group consist mainly of PI I/II, PKPI, PCI, A5-inhibitor and Bowman Birk inhibitor (BBI) [8]. Prior research has suggested that potato peptides have biological activities. For example, they have been shown to regulate blood pressure and serum cholesterol [9] and have ACE-inhibitory and antioxidant properties [10]. Potato trypsin inhibitor (PTI) fractions may also be of great interest, as they are reported to increase satiety levels, reduce food intake and delay gastric emptying by elevating Cholecystokinin levels in plasma [11,12].

Conventionally, the industrial production of the protein fraction from potato juice is achieved by applying isoelectric precipitation or heat coagulation, after the extraction of the starch. In these processes, despite the high yield achieved, modifications in the protein structure due to harsh conditions lead to decreases in functionality compared to the native isolate, resulting in ingredients mainly suitable for animal feed [13]. To overcome this challenge, technologies such as ammonium sulfate ($(NH_4)_2SO_4$) precipitation, membrane separation (micro, ultra and nano-filtration), ion exchange (IEX) and expanded bed adsorption (EBA) chromatography have been applied as more gentle approaches in the recovery and isolation of potato protein [14]. Furthermore, using these processes, based on the intrinsic physical differences between the proteins in the extract, it is possible to remove antinutritional factors such as glycoalkaloids (chaconine and solanine). These processes can result in an ingredient of high technological functionality [15]. The isolates obtained still maintain high solubility and allow the design of various structures in foods, due to their gelling, emulsifying and foaming properties.

In this study, it was hypothesized that potato protein digestibility is affected by its processing history when used as an ingredient, ultimately affecting its biological functions. For this reason, different structures were prepared using potato protein isolates and were subjected to different in vitro digestion kinetics. Although animal and human intervention trials are still the gold standard to measure protein digestibility, these tests are ethically challenging, and for this reason, in vitro methods are often preferred to provide a mechanistic understanding of the effect of the food matrix on digestion behavior. To test this hypothesis, a commercial potato protein ingredient (AVEBE, Solanic® 200) [16], with high functionality (solubility, foaming and gelling capacity) was used to prepare different matrices (suspension, gel, unstable and stable foam) of isocaloric content. These systems were then subjected to a semi-dynamic in vitro digestion study, a method that takes into consideration the structural changes occurring to food in the stomach, and provides the necessary conditions to understand the fate of the food matrix in the gastrointestinal tract. This was in order to determine how the structures formed before and during gastrointestinal transit may play a role in their digestibility.

2. Materials and Methods

2.1. Preparation of Potato Protein Structures

A commercially available, patatin-rich potato protein isolate (PoPI) (Solanic® 200, Avebe, Veendam, The Netherlands) was used in this study. The manufacturer's specification indicated that PoPI contained 87.5% protein (as is), 3.5% fiber (as is), 0.2% carbohydrates (as is), 0.2% fat (as is), 5% moisture content and 4% ash (as is). Four different structures containing the same amount of protein (4% protein, *w*/*w*) were prepared: a suspension (PoPI-S), a heat induced gel (PoPI-G), a foam (PoPI-F) and a heated foam (PoPI-HF). PoPI-S was prepared by dissolving PoPI powder in deionized water and stirring overnight at room temperature. As per manufacturer's specification, the protein showed $\geq$85% solubility. PoPI-G was instead prepared by heating the suspension at 90 °C for 15 min in a water bath. This caused the formation of a gel. PoPI-F was a foam prepared by whipping the suspension for 3 min in a Thermomix® (Vorwerk) (Wuppertal, Germany) (using the butterfly whisk at speed #3). The fourth sample (PoPI-HF) consisted of a heated foam structure produced by treating the previously mentioned foam in a microwave (Siemens®,

BF525LMS0, Aarhus, Denmark) at 360 W for 1 min. The temperature inside the heat-set foam was 72 °C, as measured immediately after heating using a thermocouple (Testo 100®, Titisee-Neustadt, Germany).

2.2. Determination of Trypsin Inhibitory Activity in PPI

The residual trypsin inhibitor (TI) activity was determined as previously published [17]. Trypsin from porcine pancreas (T4799) was dissolved in 1 mM HCl solution (pH 3.0) to a concentration of 500 μg/mL. Nα-Benzoyl-l-arginine 4-nitroanilide hydrochloride (BApNA) dissolved in Tris-buffer solution (400 μg/mL) (0.05 M, pH 8.1) was used as substrate, and 30% acetic acid (*v*/*v*) was used to stop trypsin activity. The blank was prepared by adding 1 mL of acetic acid and 2 mL of trypsin solution; to this mix, 2 mL of deionized water and 5 mL of BApNA solution were subsequently added. For the samples/control: 2 mL of trypsin solution was mixed with 2 mL of PoPI samples, or with deionized water for the control. Once mixed, the tubes were submerged in a water bath (37 °C, 10 min). Subsequently, 5 mL of BApNA substrate solution was added, and incubated for 10 min. Finally, 1 mL of 30% acetic acid was added to stop the reaction. The samples and controls and blank were transferred to 1 × 1 cm cuvettes and the absorbance of the resulting product measured 410 nm using a Cary 60 Bio UV-spectrophotometer. A decrease in 0.01 U of absorbance in the assay (15 min, 37 °C) was defined as one unit of trypsin inhibitory activity (TIA). The extract was diluted to reach TI (%) values ranging between 25–70%. The residual trypsin inhibitory activity and the residue percentage were calculated using the following equations:

$$\text{TIA (U)} = \frac{(A_c - A_0) - (A_t - A_0) \times N}{0.01}$$

$$\text{TIA (\%)} = \frac{(A_c - A_0) - (A_t - A_0)}{(A_c - A_0)} \times 100\%$$

where A_c is the absorbance of the control, A_t is the absorbance of the sample, A_0 is the absorbance of the blank, and N is the dilution factor.

2.3. Gastrointestinal Simulated Semi-Dynamic in Vitro Digestion

A standardized semi-dynamic in vitro digestion model [18] was used with the purpose of simulating the environment during gastrointestinal transit. The protein hydrolysis was carried out under physiologically relevant conditions (increasing enzyme concentrations, changes in the pH and emptying dynamics). Electrolyte simulated salivary fluid (eSSF), gastric fluid (eSGF) and simulated intestinal fluid (eSIF) were prepared according to INFOGEST 2.0 protocol, as well as their mixing ratio with food and enzymes [19]. All enzymatic activities were also measured according to the protocol. Reagents were of analytical grade.

The caloric content of the four different PoPI model foods was calculated according to the Atwater factor (4 kCal g^{-1} for protein, 4 kCal g^{-1} for carbohydrates and 9 kCal g^{-1} for lipids) based on proximate composition of the sample. The gastric emptying time was estimated based on data of in vivo emptying dynamics (2 kCal min^{-1} emptying of 500 mL food), and this time was also used to determine the rate of addition of the simulated fluids and enzymes [20]. Given the low content of carbohydrates and lipids in PoPI powder, salivary amylase and gastric lipase were omitted in the in vitro experiment. Based on the sample size (10 mL, 1.65 kcal) of the current work, the duration of gastric digestion was 41.3 min. Gastric aliquots (3.82 mL) were collected every 8.26 min, for five gastric emptying points numbered in sequence (GE1–GE5). The total amount of 0.5 M HCl needed for gastric digestion was estimated by testing the volume needed to reach a final pH ~2. Porcine pepsin (2000 U mL^{-1}) (Sigma Aldrich, 3760 V, 3300 U mg^{-1}, Darmstadt, Germany) was added at a secretion rate of 48.42 μL min^{-1}. Immediately after collection, the pH of the gastric samples was raised to 7 by adding NaOH (0.5 M), so pepsin activity was inhibited. Subsequently, the emptied gastric samples were subjected to static intestinal digestion [19,21]. The digesta, from the five emptying points (GE1–GE5), were mixed in falcon tubes for two hours at 37 °C in an incubator with eSIF with bovine bile extract (10 mM bile salts) (Sigma Aldrich, B8631, Darmstadt, Germany) and pancreatin (200 U mL^{-1}, trypsin activity in eSIF) (from porcine pancreas, Sigma Aldrich, P1750-25G, Darmstadt, Germany). To stop trypsin and chymotrypsin activity, 1.91 mL of 0.01 M Pefabloc® (Darmstadt, Germany) was added. The intestinal endpoints, indicated as IE1–IE5, correspond to the previous gastric emptying points (GE1–GE5). Samples were immediately frozen at −80 °C for 24 h and then stored at −20 °C.

2.4. *Confocal Laser Scanning Microscopy (CLSM)*

Confocal laser scanning microscopy (CLSM) (Nikon Eclipse Ti series, Nikon Instruments Inc., Amsterdam, The Netherlands) was used to observe the microstructure of the four potato protein samples emptied at the different stages of gastrointestinal digestion. Aliquots of 1.5 mL were stained with 30 μL of 1 mg/mL fluorescein-5-isothiocyanate (FITC). The pinhole diameter was maintained at 1 Airy Unit. Images were taken at 20× dry objective and 60× water immersion objective, with a refractive index of 1.0 and 1.33, respectively. A 488 nm helium/neon laser was used and the filter was set at 475–650 nm to image labelled proteins.

2.5. *Sodium Dodecyl Sulphate-Polyacrylamide Gel Electrophoresis (SDS-PAGE)*

Protein breakdown during gastric and intestinal digestion was followed using SDS-PAGE. Samples (13 μL) of undigested material, the 5 gastric and intestinal endpoints, were mixed with 5 μL of NuPAGE® LDS sample buffer and 2 μL NuPAGE reducing agent (Thermofisher, Waltham, MA, USA), and heated at 90 °C for 5 min. Samples were loaded on NuPAGE® SDS-PAGE bis-tris gels (4–12%). Running buffer was prepared by mixing 50 mL NuPAGE® MES-SDS Running Buffer (20×) with 950 mL deionized water. The separation process was run for 45 min at 200 V. Protein bands were stained with SimplyBlue (Invitrogen, Thermo Fisher Scientific, San Jose, CA, USA) and images were analyzed using ChemiDoc XRS+ and Image Lab software (Bio-Rad Lab., Hercules, CA, USA).

2.6. *Size Exclusion Chromatography*

Size exclusion chromatography (SEC-HPLC) was performed for the gastrointestinal points (GE1-GE5, I1 and I5 at 120 min) of the four different digestion treatments. The impact of the different treatments on the pattern formation of peptide aggregates in the soluble phase in different stages of gastrointestinal digestion was determined by analyzing the size distribution profile from the digesta. The analysis was carried out with a HPLC system (Agilent Technologies 1100 series) equipped with a TSKgel column G2000SWXL (7.8 mm × 600 mm) (Tosoh Corp., Merck, Germany). The digesta were centrifuged at 10,000× *g* for 30 min at room temperature, and the supernatant, corresponding to the bioaccessible fraction [22], was diluted 1:4 (sample: acetonitrile 30%, *v*/*v*) and filtered using a syringe filter (0.45 μm PVDF membrane, Gilson Scientific Ltd., Luton, UK). The separation was obtained isocratically with a mobile phase consisting of 0.1% (*v*/*v*) trifluoroacetic acid and 30% (*v*/*v*) acetonitrile, at a flow rate of 0.5 mL min^{-1}, measuring the absorbance at 214 nm. Integration of the chromatograms was obtained using Agilent Chemstation software (Agilent Technologies, Santa Clara, CA, USA). To estimate the molecular weight distribution of the peptides present in the bioaccessible fraction, their elution time was compared to the elution time of standards: bovine serum albumin (67 kDa), carbonic anhydrase (29 kDa), β-lactoglobulin (18.4 kDa), aprotinin (6.5 kDa), and the amino acids histidine-leucine (0.268 kDa), phenylalanine (0.165 kDa) and glycine (0.075 kDa).

2.7. *Degree of Hydrolysis*

The degree of hydrolysis was measured using the o-phthaldiadehyde (OPA) assay. The digested samples (0.5 mL) were mixed with Trichloroacetic acid (0.83 mL) (TCA) (3.12%

w/*w*, final concentration) to precipitate insoluble protein. Samples were then centrifuged at 10,000× *g* for 30 min at room temperature. Thereafter, the supernatant was filtered with a 0.45 μm filter (PVDF membrane, Gilson Scientific Ltd.) (Luton, United Kingdom) and mixed with a reagent mixture containing ultrapure water (88.1%, *w*/*w*), sodium tetraborate (3.81%, *w*/*w*), SDS (0.1%, *w*/*w*), Dithioerythritol (DTE) (0.088%, *w*/*w*), and OPA (0.08%, *w*/*w*), previously dissolved in 4 mL ethanol. Samples/standards (10 μL) were mixed with 250 μL of OPA reagent and the reaction was allowed to proceed for 15 min. Free NH_2 groups values were determined subsequently using a spectrophotometric assay and the absorbance was recorded at 340 nm using a plate reader. A stock solution of 2 mg/mL of L-leucine (131.17 g/mol) was used to prepare the calibration curve with ranging concentrations from 0 to 15 μM. Each measurement was performed in triplicate. The degree of hydrolysis was calculated as a ratio of the number of hydrolyzed peptide bonds divided by the number of total peptide bonds present [23].

2.8. Free Amino Acid Analysis by Triple Quadrupole Mass Spectrometry

To estimate the protein digestibility, the amount of free amino acids in the supernatant of the intestinal endpoints (I5) was calculated using a triple quadrupole mass spectrometer (TQMS). To obtain the soluble fraction, the stated procedure in SEC-HPLC section was followed. Once centrifuged and filtered, 20 μL of sample was diluted with 180 μL of 0.1 M HCl, and 50 μL of internal standard consisting of stable isotopically labelled amino acids ["Cell Free" Amino acid mix (20 AA, U-13C, 97–99%; U-15N, 97–99%) (Cambridge Isotope Laboratories, Cambridge, MA, USA)] was added to the mixture. To prepare the stock solution of the internal standard, 0.1 M HCl was used and kept in separate vials at −20 °C. The samples were then subjected to a chromatographic separation on a UHPLC system (Agilent Technologies, 1290 Infinity II), using an Intrada Amino Acid column, 150 × 3 mm, 3 μm (Imtakt, Portland, ORE, USA) in combination with a Van-Guard Pre-column ACQUITY UPLC BEH HILIC, 5.0 × 2.1 mm, 1.7 μm (Waters, Ireland), at a flow rate of 0.6 mL/min. The analytical column was thermostatted at 35 °C (±0.8 °C) and the injection volume was 5 μL. Detection and mass analysis were performed on a Triple Quadruple Mass spectrometer (TQMS) system (6495 Agilent Technologies, Santa Clara, CA, USA) using MassHunter® software (Agilent Technologies, Santa Clara, CA, USA) to interpret the data. Acetonitrile with formic acid (0.1%, *v*/*v*) and ammonium formate were used as solvents for the separation in the chromatographic column. A gradient elution program was used going from 20% (ammonium formate, 100 mM) (0 min) to 100% (12 min), followed by an isocratic elution of 100% ammonium formate for 2 min. The column was then re-equilibrated for 15 min with 20% ammonium formate after each run.

The reagents used in the mobile phase were the following: Acetonitrile, LC-MS grade (Merck, Steinheim, Germany); Ultrapure water (18.2 MΩ) (MilliQ, Millipore, Darmstadt, Germany) used in the preparation of solutions; formic acid, LC-MS grade (Sigma Aldrich, Steinheim, Germany) and Ammonium formate (Sigma Aldrich, Steinheim, Germany).

2.9. Statistical Analysis

The measurements reported for protein particle size, degree of hydrolysis, and trypsin inhibitory activity are the results of three independent measurements, while two independent measurements were used to estimate the molecular weight distribution and free amino acids. Statistical analysis was performed using IMB SPSS v22 (Armonk, IBM Corp, New York, NY, USA), and minimum significance was set at the 5% level ($p < 0.05$). One way analysis of variance (ANOVA) with Tukey HSD was applied to trypsin inhibitory activity and free amino acids and was conducted to determine the significance of the treatments, while a two-way multivariate analysis of variance (MANOVA) was applied to SEC-HPLC Mw distribution.

3. Results and Discussion

3.1. Microstructural Characterization

The visual appearance of the four structures prepared from the potato protein isolate is shown in Figure 1. Representative microstructure images corresponding to each of the four different potato protein treatments (PoPI-S, PoPI-G, PoPI-F, PoPI-HF) are presented in Figure 2. Images depict, for the same treatment, the initial samples (U), sequential stages of gastric emptying (G1, G3 and G5) and the final intestinal stage (I5). The latter represents the most extensively digested sample after simulated intestinal digestion. Figure 2 shows visible differences between the colloidal structures of the four matrices, before and throughout the in vitro digestive process. Images at higher magnification can be also found in the Supplementary Material.

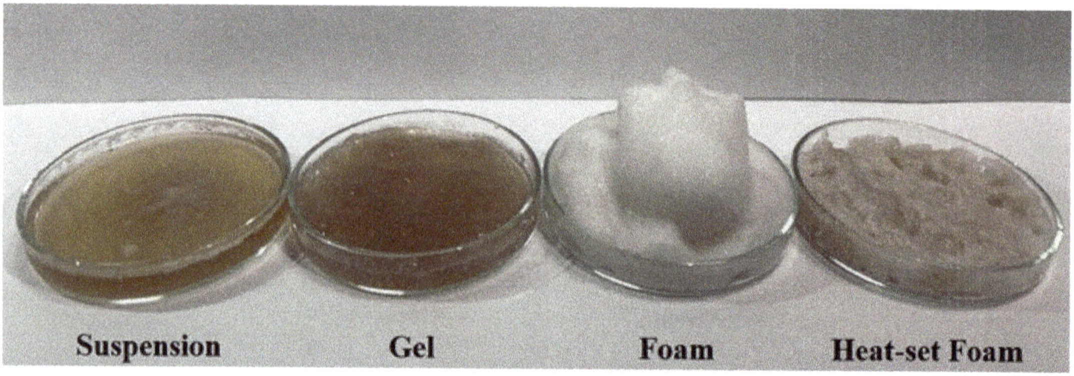

Figure 1. Visual appearance of potato protein isolate structures (4% protein, *w/w*).

The untreated dispersion (PoPI-S) was mostly homogeneous with a few small particles in suspension (Figure 2A). These particles represent residual insoluble powder sized between 10 and 50 μm. The apparent average diameter of these suspensions was about 20 μm when measured by integrated light scattering (data not shown). These results are well aligned with the known solubility of this commercial isolate. During gastric transit, the suspension showed the formation of aggregates. At the acidic pH of the gastric stages G1 and G3 (pH 5.5–3.6, Figure 2b,c), the protein suspensions were characterized by the presence of interacting particles, and at G3 there seemed to be a semi-continuous matrix, which could be attributed to the formation of an acid-induced network structure. It is indeed reported that protein-protein interactions occur as the surface charge of proteins get closer to their isoelectric point (4.5–5.1), leading to aggregation of the globular patatin [24]. At the final stage of in vitro gastric digestion (Figure 2d) at pH 2.0 (G5), the network is no longer visible. The residual aggregates, less than 50 μm in size, are diluted in the gastric juices, and are characterized by an increased charge repulsion, as the environment is now far from the isoelectric point of the protein particles [25]. The persistence of globular particles at this point would suggest an incomplete hydrolysis by pepsin at these gastric emptying times. Figure 2e depicts the microstructure of the final intestinal endpoint (I5) of PoPI-S. At this stage, the suspension shows extensive digestion, with only a few aggregates present, also deriving from the inhomogeneous suspension of the simulated in vitro intestinal juice.

Figure 2 also shows the microstructure changes occurring to the potato protein heat-induced gel (PoPI-G). Before digestion (Figure 2f), the microstructure consisted of a continuous and compact-dense network, characteristic of heat-induced protein gels. This gel, formed at neutral pH, is stabilized mainly by hydrophobic interactions [25,26], given the lack of sufficient internal disulfide bonds in the potato protein isolate. This would imply the extensive unfolding and exposure of non-polar moieties of the proteins, and make the proteins more prone to accessing the hydrolytic enzymes during digestion. Previous work has reported a denaturation temperature for patatin of 60.8 °C, while trypsin inhibitor gelation is induced after 70 °C [24,27]. In the present work, a denaturation peak for the commercial potato protein isolate was 68.7 ± 0.28, measured using differential scanning calorimetry (data not shown). Similar results were also found by Elahi and Mu [28]. During acidic gastric conditions (Figure 2g–i), the structure of the digesta transiting to the intestine still showed a compact network, with a brittle appearance and similar features between G1 and G3 emptying points (corresponding to a pH drop from 5.0–3.9). Only some large particles were visible at the final gastric point G5 (Figure 2i), and a homogeneous suspension of small aggregates was left at the final intestinal stage, I5 (Figure 2j).

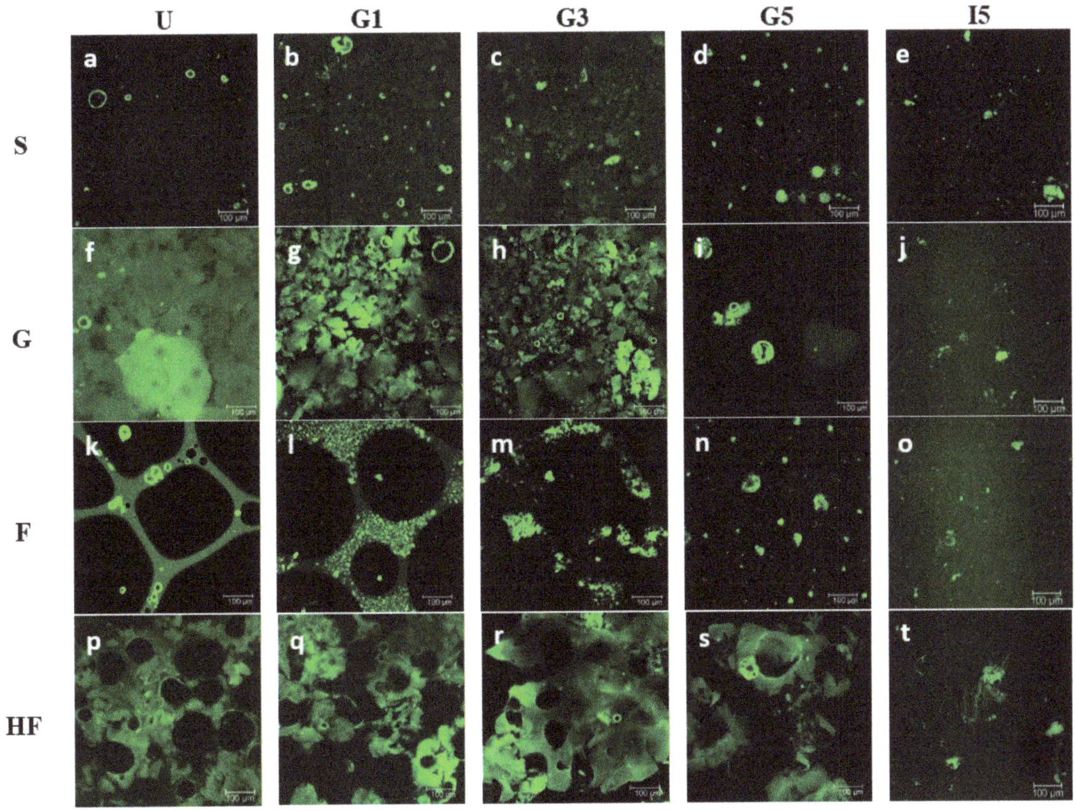

Figure 2. Representative confocal scanning laser microscopy (CSLM) images of four different potato protein isolate structures (4% protein, *w*/*w*). Structures: Suspension (S) (figures **a**–**e**), Gel (G) (figures **f**–**j**), Foam (F) (figures **k**–**o**), Heated Foam (HF) (figures **p**–**t**). Different gastric endpoints shown: GE1 (8.26 min), GE3 (24.78 min), and GE5 (41.30 min). Scale bar: 100 µm.

The initial microstructure of PoPI-F was that of a typical foam (Figure 2k), containing large gas bubbles stabilized by a protein matrix, and with lamellae containing the remaining unadsorbed protein. The foam still retained its structure, at least in part, with the decrease of pH to 5.5 (Figure 2l), although the lamellae were thicker after the first stage of gastric digestion, due to dilution with eSSF, eSGF and the lower pH. However, the air bubbles were still quite visible and the protein particles seemed to be present in the interstitial phase as individual aggregates. This may indicate a difference in the structure of the proteins due to their surface denaturation. At G3 (pH 3.5) (Figure 2m), the air bubble structures collapsed, and an acid-induced protein-protein network around the lamellae was now visible, showing remnants of a gelled lamellar phase. At G5, it was no longer possible to collect the total calculated aliquot for each gastric emptying point (3.22 mL instead of 3.82 mL) as some of the foam structures persisted in the in vitro simulated gastric vessel. It is important to note that in vivo intervention studies have suggested the potential of aerated foods to increase gastric volumes, delay gastric emptying and reduce appetite [29]. The current in vitro study supports such findings, as it similarly shows the presence of significant phase inhomogeneities during digestion of a protein foam. Nevertheless, it is important to take into consideration parameters such as rheological properties and the stability of the foam, as well as the ability of the lamellae to coalesce during digestion, to determine the impact of these types of structure on gastric emptying and appetite reduction. A higher viscosity has been associated with a reduced gastric emptying rate of isocaloric and isovolumetric meals [30]. In order to characterize the rheology of the entire digesta, it has been recommended to apply oscillatory shear stress and a large deformation steady shear [31]. By the final gastric point, G5, the microstructure of what was emptied (Figure 2n) was quite similar to that of the PoPI-S with some residual particles. In addition, the final in vitro intestinal stage (Figure 2o) showed similarities to the unheated suspension.

The heated foam (PoPI-HF, Figure 2p) also showed a unique microstructure. In this case, the structure was very different than that of the unheated PoPI-F. There was a dense continuous protein network, but with trapped air bubbles. These air bubbles still showed round or ellipsoidal structures, suggesting that the PoPi created viscoelastic interfaces, able to withstand the heating process. The microstructure of the gastric emptied fractions (Figure 2q–s) in this case was somewhat similar to that of the heated PoPI-G. The structure was quite similar between G1 and G3 (Figure 2q,r), and large aggregates were still present at the final gastric stage, G5 (Figure 2s). The microstructure of the intestinal sample, I5 (Figure 2t), also shows more residual aggregates compared to the other treatments. The amount of gastric digesta emptied at each stage of the in vitro gastric digestion was also lower than that of the PoPI-F, as in this case, only approximately 3 mL could be emptied at each gastric emptying point, demonstrating a much higher resilience of this sample to digestion and emptying. It is possible to attribute this to the rheological properties and the resistance of this heat-set foam to gastric juices when compared to the unheated PoPI-F. These results are in full agreement with recent reports [32] suggesting that hunger suppression induced by aerated foods could be prolonged by enhancing foam stability in gastric conditions, as this increases gastric volumes and would decrease gastric emptying rates. For all the samples, the microstructural behavior of the different systems characterized by microscopy during the gastric and intestinal stages in the experiment was highly reproducible.

3.2. Sodium Dodecyl Sulphate-Polyacrylamide Gel Electrophoresis (SDS-PAGE)

With the aim of understanding the differences in the extent of protein breakdown between the four different isocaloric matrices prepared with potato protein, their non-digested state (U) as well as their digesta were analyzed at each in vitro gastrointestinal point (G1–G5, I1–I5) using SDS-PAGE (Figure 3). These samples represented the state of the protein transiting to the upper intestine at various digestion times. All samples shared a similar electrophoretic pattern at the initial stage (U), and their bands were assigned as reported in the literature [26,33]. Two main bands were present, a high molecular weight (Mw) protein (migrating at approximately 130 kDa) corresponding to lipoxygenase, and a second large band migrating at about 43 kDa corresponding to patatin. The protein isolate was mostly comprised of patatin; however low intensity bands were also noted for the trypsin inhibitor (TI) protein group (migrating at about 22, 15 and 9 kDa) (with two bands at 15 and 9 kDa). There were clear differences in the protein hydrolysis during gastrointestinal transit between the unheated and heated samples (Figure 3A–D). The PoPI-S and PoPI-F (Figure 3A,C) followed a similar gastric digestion pattern: the Mw of all band fractions remained unaltered from G1–G3, which indicated resistance to proteolysis by pepsin, yet due to the pH conditions they were non-optimal for the gastric enzyme's activity [34]. However, at the two last emptying points, G4 and G5, a much higher extent of hydrolysis was noted, with only traces of patatin present, as well as some products of hydrolysis migrating at low Mw. At these two emptying points, conditions were ideal (highest pepsin concentration and pH ~2.0). At G5, a clear band appeared at 37 kDa, corresponding to pepsin, as the enzyme had reached high concentrations in the gastric digesta. It is important to note that when comparing G5 for PoPI-S and PoPI-F, there were some differences in the polypeptide distribution, with larger fragments left in the suspension, compared to the PoPI-F foam samples, and higher amounts of trypsin inhibitors still intact. Figure 3A,C also show the various stages of emptying after intestinal digestion (I1-I5) for PoPI-S and PoPI-F. There seemed to be more undigested protein in the suspension compared to the foam (Figure 3A,C), and the trypsin inhibitor bands were still visible at 15 kDa, in both suspension p0 and foam samples at the early stages of gastric emptying.

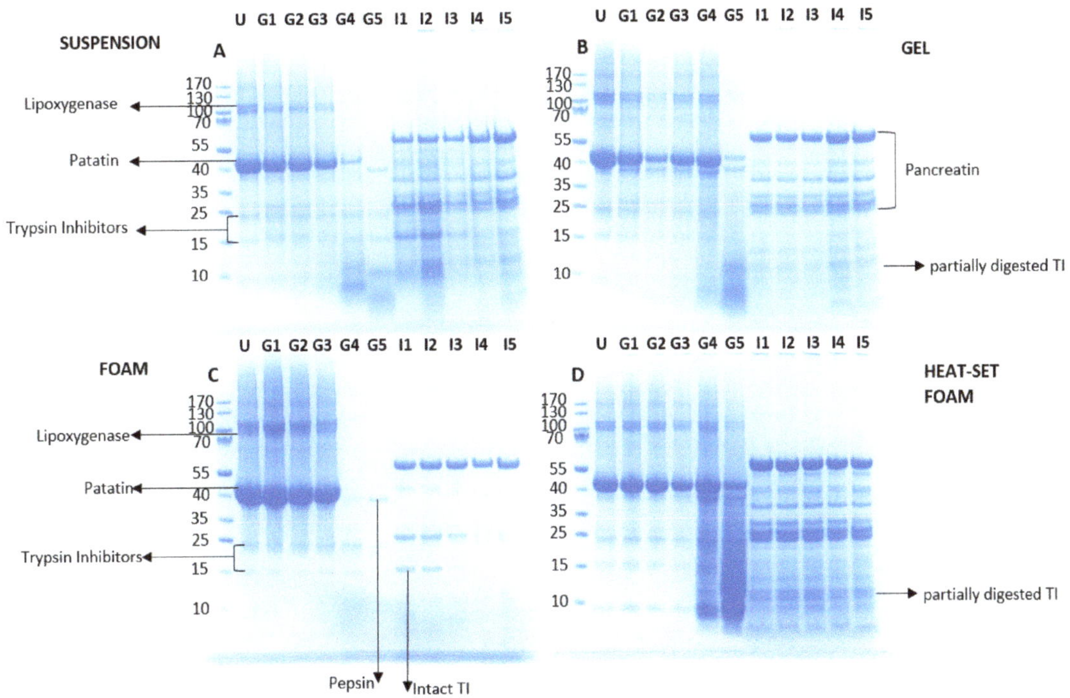

Figure 3. SDS-PAGE profiles of four different potato protein isolate matrices (4% protein, *w*/*w*) under reducing conditions: suspension (**A**), gel (**B**), foam (**C**), heat-stable foam (**D**). Undigested (U), and at the following gastric emptying points: G1 (8.26 min), G2 (16.52), G3 (24.78 min), G4 (33.04 min), G5 (41.30 min). On the right-hand side of each gel, the samples loaded are the corresponding intestinal points of each gastric point (I1–I5). Molecular weight standards are indicated on the left side, and he tmain protein fractions identified are indicated with arrows.

In the PoPI-G samples (Figure 3B), the protein seemed to be more resistant to digestion, with both lipoxygenase and patatin still visible at G4, and also at the final gastric stage (G5). Furthermore, in the PoPI-G samples, a new band (37 kDa) migrating under the native patatin fraction (43 kDa) appeared during gastric digestion G1–G4. This cannot be attributed solely to pepsin, as its concentration was not high enough to be detected at these early stages [35], and it may suggest that in the heated samples, patatin is subjected to a partial hydrolysis. As opposed to the two unheated PoPI structures (S, F) (Figure 3A,C), the heated gel still showed residual patatin at G5, as well as fragments migrating at low Mw. Previous studies have shown the resistance of native proteins to gastric digestion when compared to their denatured counterparts by using whey as a model protein [36,37]. This was attributed to peptide bonds in the native structure of whey proteins hindering the accessibility of the proteases [38]. This is not the case for potato protein isolate, where there was an apparent higher resistance to hydrolysis under gastric conditions compared to the unheated suspension. This may be attributed to the heat-induced denaturation of the protein, and/or to the physical structures formed in the gels. It has been previously suggested [39] that gel structures may provide a less effective medium for pepsin diffusion. The TI was present during the entire gastric digestion, and partly degraded under intestinal conditions, as shown by its migration at a lower Mw in Figure 3B.

The fractions with the largest peptide fragments transiting to the intestinal phase were noted for the heated foam. Figure 3D illustrates the protein distribution during in vitro digestion of the PoPI-HF. This treatment, as already suggested by the microstructural observations in Figure 2, showed the highest resistance to hydrolysis, during both the gastric and intestinal phase, demonstrating its persistence during gastrointestinal transit due to the structures created. In the initial stages of gastric digestion, patatin did not show changes, nor the appearance of the second band at lower molecular weight, which is shown in Figure 3B (PoPI-G). However, similarly to the gel, the patatin band was disrupted in the heat-set gel at the late stages of gastric digestion, with fragmentation of patatin visible only at G4 and G5, possibly due to a delay in digestion kinetics. Both PoPI-G and PoPI-HF showed a persistent patatin band at G5. The PoPI-HF also showed the largest population of partly degraded protein and TI during the latest stages of gastric digestion and in the intestinal stages. In particular, partly digested TI remained constant at all intestinal points (Figure 3D).

3.3. Determination of Trypsin Inhibitory Activity in PPI/Processing Effect on Trypsin Inhibitory Activity in Potato Protein Isolates

The residual trypsin inhibitory activity (TIA) present in the various PoPI processed samples was estimated (Table 1). All samples still showed residual trypsin inhibitory activity. The results support the findings described in Figure 3, in which TI bands were present in the undigested samples. The PoPI dispersion (72.8% ± 1.1) had the highest TIA, fully in agreement with the native character of the isolate. As expected, the unheated foam, PoPI-F (67.9% ± 2.9), also showed a high TIA, and no significant difference compared to the dispersion. Despite the drastic change in food structure, there was no surface denaturation of the TI. Heating treatment, on the contrary, showed a decrease in TIA; PoPI-G showed a slight reduction to 62% of TIA, and in the case of PoPI-HF, the reduction was significantly lowered to 25.6%. It has been previously shown that the structural change in protein natural conformation caused by heat-induced denaturation produced a non-reversible change, leading to a loss of inhibitory activity [40–44].

Table 1. Trypsin Inhibitory Activity (TIA, Units/mg solids) and percentage of inhibition (TIA%) (see methods) for various treatments. One unit of TIA produced 50% inhibition of 2 Trypsin units. Values are reported as the average of three replicates with standard deviation. Different letters indicate statistically significant differences between treatments.

Trypsin Inhibitory Activity	PoPI-S	PoPI-G	PoPI-F	PoPI-HF
TI (U/mg PoPI)	3.1 ± 0.0 [a]	2.7 ± 0.0 [b]	2.9 ± 0.1 [a]	1.1 ± 0.0 [c]
TIA (%)	72.8 ± 1.1 [a]	61.9 ± 1.2 [b]	67.9 ± 2.9 [a]	25.6 ± 0.3 [c]

3.4. Peptide Size Distribution and Degree of Hydrolysis during Gastric Emptying and Intestinal Digestion

To follow the distribution of peptides formed during digestion depending on the matrix structure, the molecular weight distribution of the peptides present in the bioaccessible phase (centrifugal supernatants) was analyzed by size exclusion chromatography, for different gastric times (G1, G3, G5) as well as two intestinal stages (I1 and I5). Representative chromatograms are shown in Figure 4. I1 and I5 represent the simulated intestinal digestion of the first and last gastric emptying time (G1 and G5). The elution profile was separated into three different peptide populations, between 80 and 30 kDa, between 30 and 1 kDa, and <1 kDa, and the percentage over total elution area is shown in Figure 5. With the exception of the heat-set foam (Figure 5d), all the other PoPI treatments showed similar trends. The gastric digesta emptied at first had large Mw peptides (80–30 kDa), and this population then decreased; the small Mw peptides (<1 kDa), on the other hand, increased, reaching values larger than 80% of the total area during the intestinal digestion. This demonstrated that the supernatant fraction in the intestinal stage contained mostly bioaccessible peptides.

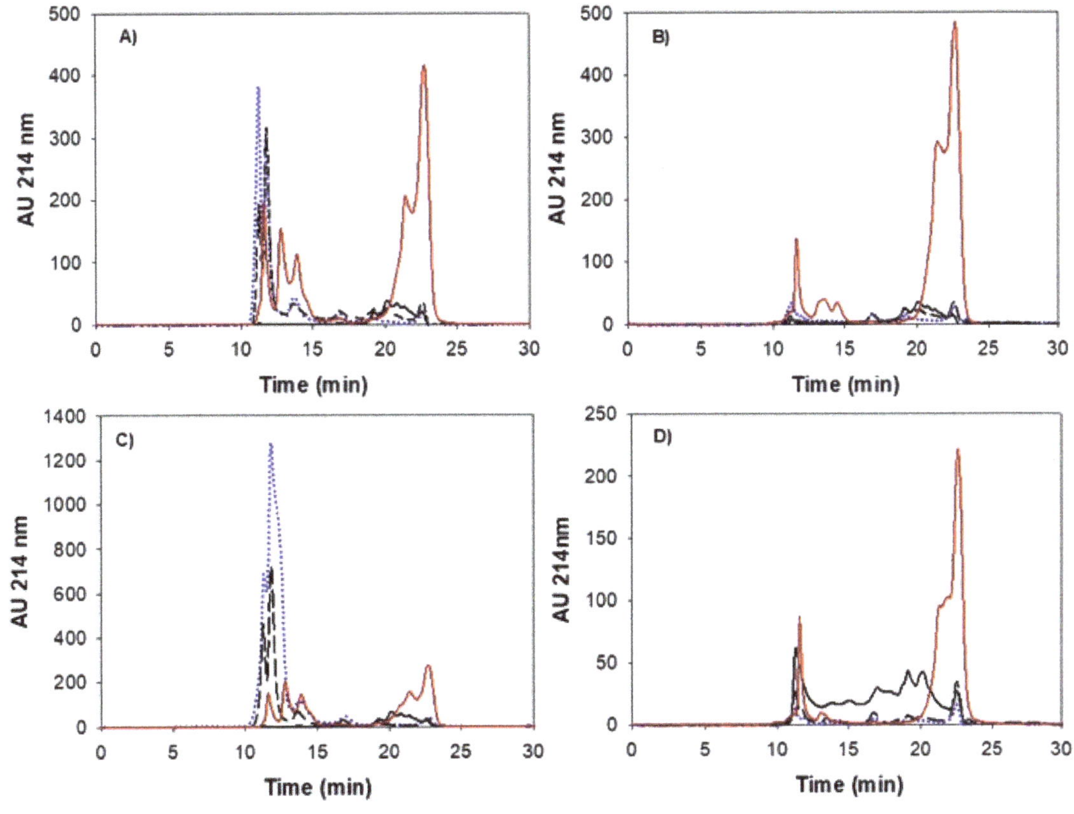

Figure 4. Representative elution patterns for supernatants of the digestates at G1, G3, G5 and I1 analyzed by size exclusion chromatography (see methods) for potato protein isolates: suspension (**A**), gel (**B**), foam (**C**), heated foam (**D**). Standard peaks elution: 100 kDa–10.6 min, 30 kDa–13 min, 1 kDa–19 min. G1 (dotted blue line), G3 (broken line), G5 (continuous black line) and intestinal, corresponding to G1 (red line).

The Mw distribution of PoPI-S extracted in the soluble fraction at the early stages of gastric digestion consisted mostly of 80–30 kDa proteins, but this population no longer existed at the end of the gastric stage (G5) (Figure 5a). The proteolysis intensified with increasing pepsin concentration (constant addition, 48.42 μL min^{-1}) and activity (due to the increasingly acidic environment), resulting in a complete shift of peptide distribution from G1 to G5, in which the 80–30 kDa group changed from 75.1% ± 1.6 to 1.9% ± 1.3, and the <1 kDa group changed from 4.0% ± 0.4 to 80.6% ± 7.4, making these peptides bioaccessible. It is important to note that even in the case of G1, the samples were fully digested once in the intestinal phase, as shown by I1.

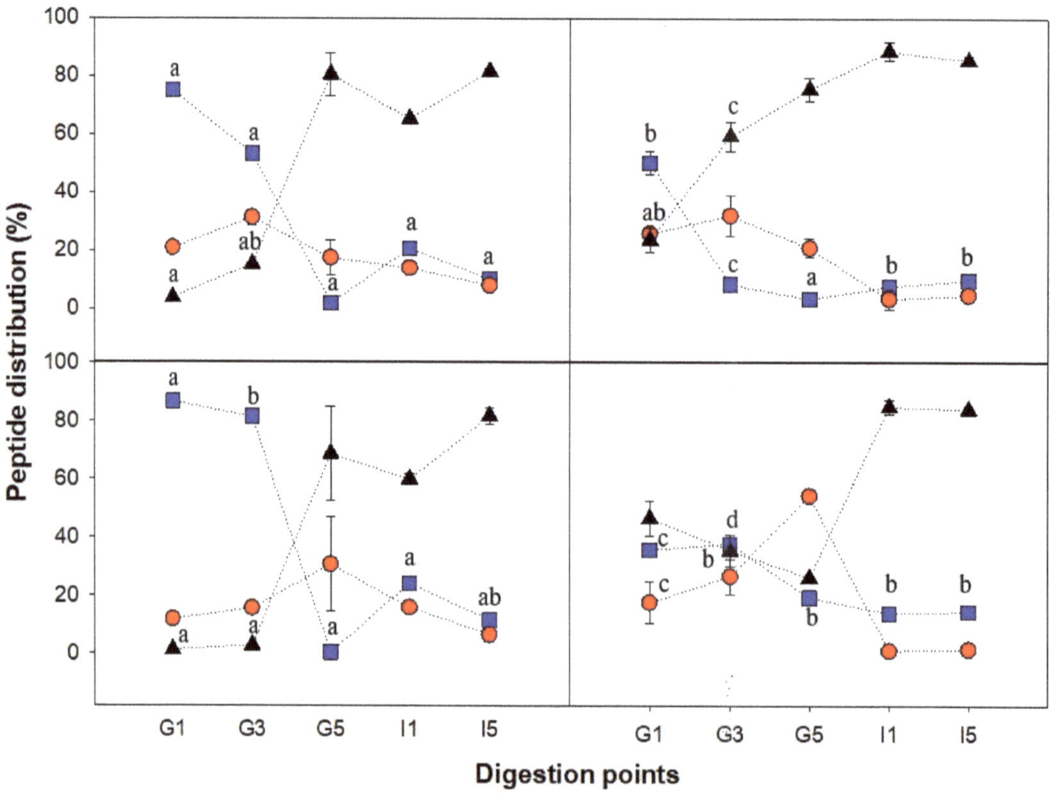

Figure 5. Molecular weight distribution of the peptides present in the soluble fraction of the four potato protein isolate (PoPI) samples. Suspension (a), gel (b), foam (c), heat-set foam (d). Three different Mw ranges are presented: 80–30 kDa (squares), 30–1 kDa (circles), <1 kDa (triangles). Bars represent standard deviation. Letters indicate differences within the same Mw population. Lines are drawn to guide the eye. Gastric stages: G1 (8.26 min), G3 (24.78 min), G5 (41.30 min), and corresponding intestinal points (I1 and I5).

When comparing the peptide population in the soluble extract between PoPI-S and PoPI-G, the heat-treated counterpart, it was clearly shown that the high Mw peptides in PoPI-G were significantly lower ($p < 0.05$) in G1, by 25%, and in G3, by 45%, while the low Mw peptide group was significantly higher ($p < 0.05$) in G1, by 29%, and in G3, by 44%. This would indicate that the breakdown of proteins was faster in the heated gel structure, compared to the unheated suspension. However, it is important to note that these were the peptides present in the soluble fraction. Indeed, the SDS-PAGE data (Figure 3) suggests a higher resistance to digestion of the proteins in the heated PoPi-G and PoPi-HF.

As already shown in Figure 3, PoPI-F followed a proteolysis trend similar to that shown in all gastrointestinal stages of PoPI-S. However, there was a higher percentage of the 80–30 kDa group in PoPI-F than PoPI-S at the G1 and G3 points. No significant differences were found in the other stages. Eventually, foam decay by liquid drainage and further collapse of the air bubbles (coalescence and Ostwald ripening) occurred, which led to a similar digestive behavior between the suspension and the foam. Both structures are examples that confirm the synergistic interaction of pepsin and trypsin to hydrolyze protein. A study by Rivera del Rio and colleagues [45] previously demonstrated the increased efficiency in hydrolysis when both enzymes were present under ideal conditions, signaling that gastric hydrolysis facilitated the trypsin hydrolysis of a model protein (Bovine

Serum Albumin) in the small intestine. The increase in small peptide distribution (<1 kDa) from I1 (59.9% ± 1.8) to I5 (81.9% ± 2.9) in PoPI-F as well as a similar increase in the same group for PoPI-S confirm the previous findings.

In the case of PoPI-HF, very little high Mw peptides were present in the bioaccessible fraction, even at G1. This may indicate a very low rate of dissolution of the protein at the initial stages of digestion. At G1, the peptide distribution pattern of this food structure already consisted mostly of low Mw peptides (46.5% ± 6.0). Despite the continuous addition of pepsin and gastric juices, the low Mw peptide group decreased continuously in G3 and G5 (35.5% ± 5.3 and 26.2% ± 0.8, respectively). The slow protein breakdown from G1 to G5 was attributed to the high stability and resistance of the heat-set foam, as the structure may have hindered the diffusion of gastric juices in the vessel. This resulted in a delayed and heterogeneous digestion process, which resulted in limited gastric emptying. This experimental set-up was considered relevant to in vivo conditions as HCl is released from parietal cells, which are present in the glands within the fundus and body of the stomach (from the top) [46], while mixing intensifies in the antrum [39]. Despite the structure resistance to gastric breakdown and slow gastric emptying, once the suspensions were transferred to the intestinal stage, the digestion was immediate. Both fractions emptied at G1 and G5 showed a high release of low Mw peptides after the intestinal stage (I1, I5), again demonstrating the prompt digestion of the potato protein isolates.

To further determine the impact of pepsin and trypsin during the gastrointestinal points of each PoPI structure, as well as the effect of enzymatic interaction, the degree of hydrolysis (DH%) was studied. The extent of protein hydrolysis in all the gastrointestinal digestion points was evaluated by measuring the free amino groups released and expressed as g of L-Leucine equivalents per total amount of protein in the initial sample (40 g/L), as shown in Figure 6. All samples showed a very low degree of hydrolysis (DH) (less than 10%) regardless of gastric emptying times. However, the samples showed a higher degree of hydrolysis after intestinal digestion (≥51.2%). DH was slightly lower in I1 compared to the later points, confirming that an increase in pepsin concentration and gastric incubation time ultimately favored trypsin hydrolysis. Differences in DH kinetics were marked by the matrix structure of the heated foams, which showed lower values compared to the PoPI-G, also heated.

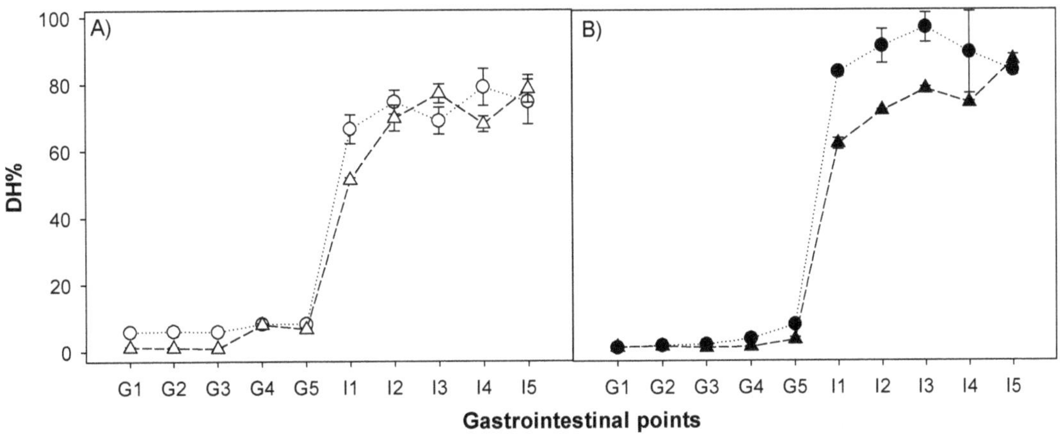

Figure 6. Release of free amino groups in the different potato protein isolate treatments. (**A**) Unheated treatments, suspension (PoPI-S, empty circle), foam (PoPI-F, empty triangle); (**B**) heated treatments, gel (PoPI-G, filled circle), heat-set foam (PoPI-HF, filled triangle). Gastric emptying points: G1, G2, G3, G4, G5 and intestinal points corresponding each gastric point (I1, I2, I3, I4 I5). Dotted lines are used to guide the eye.

Among all the samples, the suspension had the highest release of free NH_2 groups (5.8% ± 0.3) under early gastric conditions (G1), while at this same time point, the other treatments presented values ~1%. These values slightly increased at the later gastric stages. At the last gastric emptying point, the suspension, gel and foam had similar DH values (~8%), while PoPI-HF only reached 3.2% ± 0.8. Thus, the differences in gastric digestion between the four PoPI structures could be due to the influence of matrix structure complexity on pepsin diffusivity in the sample [47].

As expected, all samples showed a drastic increase in DH from gastric to their respective intestinal points due to efficient activity by trypsin and chymotrypsin. As shown in Table 1, there were trypsin inhibitor activities for all samples, with the lowest residual activity being that of the heated foams. However, it is important to note that the levels of TIA were estimated to range from 125 U (in the sample) in the PoPI-S to 44 U in the PoPI-HF, and were much lower than the amount of trypsin present in the simulated intestinal fluid (764 U). Despite having a lower TIA, PoPI-HF did not show a higher DH than PoPI-G, confirming the hypothesis that the structuring in the gut influenced protein digestibility.

Remarkable differences between some of the PoPI structures appeared not only at the same point (e.g., suspension vs. gel), but also between the same structure at different intestinal points (e.g., I1 vs. I5). For instance, the suspension (PoPI-S) at I1 (66.5% ± 4.3) had a significantly higher DH than the unheated foam (51.2% ± 0.6) (Figure 4A). Nevertheless, both structures showed similar values from I2 to I5, reaching a final DH of about 75%. The released amount of free NH_2 groups by these two unheated treatments, regardless of the structure, showed an increase at the initial stages of gastric emptying, confirming the essential role of pepsin activity in increasing digestibility in the intestinal stages. It is also clear that the presence of native trypsin inhibitor did not have an impact on the hydrolysis of these samples.

Figure 6B shows the DH for the heated samples, the gel and the heated foam. PoPI-G reached the highest DH values during the simulated intestinal stage. This treatment showed fast hydrolysis kinetics, as even for the first gastric emptying stage, the values already reached 83.4% ± 1.3, and continued to grow to 96.6% ± 4.4. Despite reporting the lowest TIA, PoPI-HF had a similar DH to those of the unheated suspension. Although the gastric emptying was inhibited by the heated foam structure, once the heated protein passed to the intestinal stage, there was an increasingly higher level of DH for each gastric step, once again confirming the important role of pepsin in imparting the digestibility of the samples.

3.5. Free Amino Acid Analysis by Triple Quadrupole Mass Spectrometry

Individual free amino acids (FAA) released were studied using LC/TQMS at the first (I1) and last intestinal point (I5) to evaluate differences in the composition of the bioaccessible fractions for the four different potato matrices. The FAA values at I5 are reported in Table 2.

PoPI-S showed the highest release of FAA (Table 2), while the other three samples had similar values. This is a consequence of structural differences in the matrix, as PoPI-S rapidly releases FAA, due to its high solubility. There was no difference in the ratio of individual FAA released and values were similar among all the structures, indicating a similar digestion of the protein fractions within the treatments. Methionine was the essential amino acid with the lowest release at the last intestinal point for all PoPI matrices, followed by arginine. The limiting amino acid in potato protein sources is often considered to be methionine [48] while leucine is the limiting factor for colored potatoes [6,49]. The results of the current study would point to methionine and valine as the limiting amino acids in the commercial potato protein isolate.

Table 2. Free amino acid values obtained by LC/TQ-MS measured after in vitro intestinal digestion (I5). Unheated suspension (PoPI-S), gel (PoPI-G), foam (PoPI-F), heat-set foam (PoPI-HF). The values are expressed as mg of free amino acid per g soluble potato protein in the intestinal point (I5). Values are reported as average of two replicates with standard deviation. The amount of total free amino acids (TFAA) is also shown: total free amino acids. Values highlighted in bold correspond to limiting amino acids (lysine and histidine). Different letters indicate statistically significant differences between treatments for a given amino acid.

mg Free Amino Acids/g Soluble Protein	PoPI-S	PoPI-F	PoPI-G	PoPI-HF
	Essential amino acids			
Valine	33.5 ± 0.1 [a]	33.3 ± 7.3 [a]	24.2 ± 1.1 [a]	23.4 ± 0.5 [a]
Leucine	78.5 ± 2.4 [a]	46.8 ± 15.8 [a]	55.5 ± 0.2 [a]	58.5 ± 2.4 [a]
Isoleucine	119.1 ± 1.3 [a]	81.2 ± 28.0 [a]	97.8 ± 3.6 [a]	96.0 ± 2.5 [a]
Methionine	9.9 ± 0.1 [a]	6.0 ± 2.0 [a]	6.9 ± 0.05 [a]	7.4 ± 0.16 [a]
Phenylalanine	43.3 ± 2.1 [b]	25.5 ± 8.0 [a]	38.9 ± 1.2 [ab]	28.8 ± 2.2 [ab]
Threonine	30.3 ± 0.5 [a]	20.6 ± 7.5 [a]	23.5 ± 1.13 [a]	23.3 ± 0.3 [a]
Lysine	92.5 ± 2.1 [b]	56.1 ± 19.7 [a]	66.8 ± 1.6 [a]	67.6 ± 1.7 [a]
Histidine	45.4 ± 1.77 [b]	32.2 ± 11.7 [a]	31.7 ± 2.2 [a]	36.8 ± 3.2 [a]
Tryptophan	35.7 ± 0.9 [a]	23.5 ± 7.3 [a]	31.6 ± 0.8 [a]	27.2 ± 1.2 [a]
	Non-essential amino acids			
Tyrosine	30.4 ± 0.2 [a]	19.7 ± 6.3 [a]	26.3 ± 0.4 [a]	26.0 ± 0.9 [a]
Asparagine	17.2 ± 0.4 [a]	41.8 ± 6.2 [b]	21.4 ± 0.3 [a]	27.6 ± 2.3 [ab]
Cysteine	10.0 ± 0.1 [ab]	13.5 ± 0.4 [c]	10.4 ± 0.2 [b]	9.1 ± 0.07 [a]
Proline	2.48 ± 0.0 [a]	1.74 ± 0.7 [a]	1.6 ± 0.03 [a]	1.4 ± 0.06 [a]
Glycine	17.1 ± 4.1 [a]	25.6 ± 3.2 [a]	15.8 ± 0.6 [a]	16.7 ± 0.6 [a]
Glutamic acid	69.7 ± 0.2 [a]	38.7 ± 16.0 [a]	47.6 ± 0.2 [a]	44.4 ± 2.0 [a]
Aspartic acid	172.2 ± 12.6 [a]	102.8 ± 42.2 [a]	119.5 ± 1.3 [a]	116.1 ± 7.2 [a]
Alanine	58.2 ± 0.8 [a]	35.0 ± 11.4 [a]	43.9 ± 0.4 [a]	40.8 ± 0.2 [a]
Serine	15.4 ± 1.3 [a]	10.4 ± 4.9 [a]	14.18 ± 1.1 [a]	11.9 ± 1.0 [a]
Glutamine	64.1 ± 3.7 [a]	46.1 ± 14.6 [a]	45.7 ± 2.3 [a]	42.6 ± 4.4 [a]
Arginine	0.1 ± 0.0 [a]	0.0 ± 0.0 [a]	0.0 ± 0.0 [a]	0.0 ± 0.0 [a]
TFAA	1000 ± 35 [b]	695 ± 224 [a]	762 ± 20 [a]	743 ± 34 [a]

Gorissen and collaborators [1] reported a high level of BCAA in the composition of a commercially available potato protein source. The present study confirmed these results by demonstrating that PoPI released high amounts of branched chain amino acids (BCAA: valine, leucine and isoleucine) under intestinal conditions, amino acids responsible for muscle protein synthesis, fatigue recovery and decrease in exercise-induced muscle damage [50,51]. In our study, the values of released FAA were also high for lysine, phenylalanine, threonine and tryptophan.

Different studies have shown the high quality of potato protein, reporting high protein digestible corrected amino acid scores and digestible indispensable amino acid scores (PDCAAS = 1.0, DIAAS = 0.94, 1.00) [52–54], as it contains all essential amino acids in sufficient quantities [54]. However, the recently claimed DIAAS for potato protein reported in the literature was based on the initial amino acid composition of the product and not on the ileal bioaccessible fraction. This is of critical importance, as protein digestibility is overlooked or assumed to be 100%, which results in a predictive DIAAS value. This work calculates the DIAAS value in vitro after digestion. Based on the free amino acids quantified from the potato protein isolate (suspension) at the intestinal endpoints, I5, using the in vitro semi-dynamic INFOGEST system, an in vitro DIAAS was calculated in our study and reported, as shown in Table 3.

Table 3. Digestible indispensable amino acid score (DIAAS) calculations for potato protein based on initial amino acid composition of the potato protein sample [53] and DIAAS based on the in vitro digestibility by measuring presence of essential individual essential amino acids per g of protein (mg/g) in the bioaccessible fraction available at the ileum of the last intestinal point (I5). Free amino acids were measured by LC-TQ-MS. * Values of the reference pattern correspond to the adult maintenance amino acid pattern (mg/g) as recommended in the dietary protein quality evaluation in human nutrition report (FAO, 2013).

Essential Amino Acids	Amino Acid Content in the Reference Pattern Values (mg/g) *	Potato Protein Isolate Free Amino Acids Released Ileum (mg/g). Experimental Data	Theoretically Predicted DIAAS Based on Initial Amino Acid Composition of the Source (Before Digestion) [53]	Calculated DIAAS Based on INFOGEST In Vitro Experimental Bioaccesible Values (In Vitro DIAAS Adults)
Valine	39	34	1.38	0.87
Leucine	59	78.5	1.43	1.33
Isoleucine	30	119.12	1.56	3.97
Met + Cys	22	20	1.15	0.91
Phe + Tyr	38	73.7	2.10	1.93
Lysine	45	92.5	1.22	2.05
Threonine	25	30.4	1.65	1.22
Tryptophan	6.6	35.7	1.28	5.41

4. Conclusions

In this study, semi-dynamic in vitro digestion proved to be a useful system that not only provided information about the effect of ideal gastrointestinal conditions on protein bioaccesibility, as would the static in vitro system, but also allowed a mechanistic understanding of the effect of structural differences on the entire gastrointestinal digestive process. The obtained results using this system suggest that food protein matrices with isocaloric content could present differences in gastric emptying dynamics when considerable structural differences occur. Thus, a greater structural complexity could lead to more delayed gastric emptying, impacting the dynamics of gastrointestinal digestion. These results need to be further tested under in vivo conditions. The use of MRI could provide better insights and allow for validation of the in vitro semi-dynamic model.

Electrophoretic patterns showed visual differences in protein breakdown, especially in patatin bands during gastric points and in trypsin inhibitors under intestinal conditions. Trypsin inhibitory activity was greatly reduced by heat treatments; nevertheless, there was no substantial increase in the amount of free amino groups released as measured by OPA assay, when compared to non-heat-treated potato protein structures. This is mainly due to the low presence of TI in commercial potato protein ingredients. Lastly, an in vitro DIAAS for the potato protein isolate was calculated and the result indicates that the product is a high quality protein (DIAAS = 87). More in vivo research is still needed in order to determine the effect of consuming low amounts of trypsin inhibitors in the long term. Due to its denatured state, the gel structure not only had the highest degree of hydrolysis but the highest release of free amino groups in the early stages of intestinal digestion. Due to the easy gastric emptying, the suspension had a high release of free amino acids, but not the highest degree of hydrolysis. Contrarily, more complex structures in the heat-set foam had a higher degree of hydrolysis in the last intestinal point but a lower release of free amino acids. Thus, this study highlights the importance of applying both degree of hydrolysis by the OPA method and free amino acids by LC/TQMS in order to deepen understanding of the effect of complex structures on protein digestive mechanisms.

It is possible that this type of protein ingredient could find applications as a "slow protein" and have an impact on the modulation of postprandial metabolism. The incorporation of this ingredient in food matrices could favor the development of certain dietetic foods with low caloric contents. This work not only explored the effect of the structuring of potato protein isolates on the gut and its impact on hydrolysis kinetics, but also highlighted the importance of adequate processing conditions of refined ingredients obtained from by-products of alternative protein sources, as this highly functional ingredient has the potential to be successfully used in novel food matrices.

Supplementary Materials: The following supporting information can be downloaded at: https://www.mdpi.com/article/10.3390/nu14122505/s1. Figure S1. Confocal scanning laser microscopy (CSLM) of four different potato protein isolate structures (4% protein, *w*/*w*). Different gastric endpoints shown: GE1 (8.3 min), GE3 (24.8 min), and GE5 (41.3 min). Scale bar: 50 μm. Figure S2: Amount of soluble protein and free amino groups (measured by OPA) for G1 and G5 fractions. A: unheated treatments, suspension (PoPI-S, empty circle), foam (PoPI-F, empty triangle); B: heated treatments, gel (PoPI-G, filled circle), heat-set foam (PoPI-HF, filled triangle). Lines are drawn to guide the eyes, arrows point from G1 to G5. Figure S3. Amount of soluble protein and free amino groups for I1 and I5 fractions, corresponding to the A: unheated treatments, suspension (PoPI-S, empty circle), foam (PoPI-F, empty triangle); B: heated treatments, gel (PoPI-G, filled circle), heat-set foam (PoPI-HF, filled triangle). Lines are drawn to guide the eyes, arrows point from G1 to G5. Figure S4. Principal component analysis (PCA) carried out with all free amino acids (see Table 1). Green circles indicate free amino acid values. Values chosen for the analysis correspond to in vitro intestinal digestion points: I1 (filled in black) and I5 (empty symbol). Potato protein isolate (PoPI) structures are represented with different symbols: Suspension (S, square), foam (F, diamond), gel (G, triangle), heat-set foam (HF, hexagon).

Author Contributions: Conceptualization, L.J.-M. and M.C.; methodology, L.J.-M. and E.D.T.; formal analysis, L.J.-M. and M.C.; investigation, L.J.-M.; resources, M.C.; data curation, L.J.-M. and M.C.; writing—original draft preparation, L.J.-M.; writing—review and editing, M.C., E.D.T. and L.J.-M.; visualization, L.J.-M. and M.C.; supervision, M.C.; project administration, M.C.; funding acquisition, M.C. All authors have read and agreed to the published version of the manuscript.

Funding: The research was partly funded by the center for innovative foods at Aarhus University and the Villum Investigator program of the Velux Foundation. Project number 37759.

Conflicts of Interest: The authors declare no conflict of interest.

Disclaimer: This manuscript does not represent or intend to represent EFSA's (European Food Safety Authority) opinion.

References

1. Gorissen, S.H.M.; Crombag, J.J.R.; Senden, J.M.G.; Waterval, W.A.H.; Bierau, J.; Verdijk, L.B.; van Loon, L.J.C. Protein content and amino acid composition of commercially available plant-based protein isolates. *Amino Acids* **2018**, *50*, 1685–1695. [CrossRef] [PubMed]
2. Jiménez-Munoz, L.M.; Tavares, G.M.; Corredig, M. Design future foods using plant protein blends for best nutritional and technological functionality. *Trends Food Sci. Technol.* **2021**, *113*, 139–150. [CrossRef]
3. Drulyte, D.; Orlien, V. The effect of processing on digestion of legume proteins. *Foods* **2019**, *8*, 224. [CrossRef] [PubMed]
4. Zahir, M.; Fogliano, V.; Capuano, E. Food matrix and processing modulate in vitro protein digestibility in soybeans. *Food Funct.* **2018**, *9*, 6326. [CrossRef]
5. Korhonen, H.; Pihlant-Leppäla, A.; Rantamäki, P.; Tupasela, T. Impact of processing on bioactive proteins and peptides. *Trends Food Sci. Technol.* **1998**, *9*, 307–319. [CrossRef]
6. Kowalczewski, P.; Olejnik, A.; Białas, W.; Rybicka, I.; Zielińska-Dawidziak, M.; Siger, A.; Kubiak, P.; Lewandowicz, G. The nutritional value and biological activity of concentrated protein fraction of potato juice. *Nutrients* **2019**, *11*, 1523. [CrossRef]
7. Ralet, M.C.; Guéguen, J. Fractionation of potato proteins: Solubility, thermal coagulation and emulsifying properties. *LWT* **2000**, *33*, 380–387. [CrossRef]
8. Li, Q.; Huang, L.; Luo, Z.; Tamer, T.M. Stability of trypsin inhibitor isolated from potato fruit juice against pH and heating treatment and in vitro gastrointestinal digestion. *Food Chem.* **2020**, *328*, 127152. [CrossRef]
9. Liyanage, R.; Han, K.-H.; Watanabe, S.; Shimada, K.-I.; Sekikawa, M.; Ohba, K.; Tokuji, Y.; Ohnishi, M.; Shibayama, S.; Nakamori, T.; et al. Potato and soy peptide diets modulate lipid metabolism in rats. *Biosci. Biotechnol. Biochem.* **2008**, *72*, 943–950. [CrossRef]
10. Pihlanto, A.; Akkanen, S.; Korhonen, H.J. ACE-Inhibitory and antioxidant properties of potato (*Solanum tuberosum*). *Food Chem.* **2008**, *109*, 104–112. [CrossRef]
11. Komarnytsky, S.; Cook, A.; Raskin, I. Potato protease inhibitors inhibit food intake and increase circulating cholecystokinin levels by a trypsin-dependent mechanism. *Int. J. Obes.* **2011**, *35*, 236–243. [CrossRef] [PubMed]

12. Schwartz, J.G.; Guan, D.; Green, G.M.; Phillips, W.T. Treatment with oral proteinase inhibitor slows gastric emptying and acutely reduces glucose and insulin levels after a liquid meal in type II diabetic patients. *Diabetes Care* **1994**, *17*, 255–262. [CrossRef] [PubMed]
13. Zhang, D.; Mu, T.; Sun, H.; Chen, J.; Zhang, M. Comparative study of potato protein concentrates extracted using ammonium sulfate and isoelectric precipitation. *Int. J. Food Prop.* **2017**, *20*, 2113–2127. [CrossRef]
14. Hussain, M.; Qayum, A.; Xiuxiu, Z.; Liu, L.; Hussain, K.; Yue, P.; Yue, S.; Koko, M.Y.; Hussain, A.; Li, X. Potato protein: An emerging source of high quality and allergy free protein, and its possible future based products. *Food Res. Int.* **2021**, *148*, 110583. [CrossRef]
15. Schoenbeck, I.; Graf, A.; Leuthold, M.; Pastor, A.; Beutel, S.; Scheper, T. Purification of high value proteins from particle containing potato fruit juice via direct capture membrane adsorption chromatography. *J. Biotechnol.* **2013**, *168*, 693–700. [CrossRef]
16. AVEBE. Potato Protein Isolates. Google Patents WO2014011042A1, 16 January 2014. Available online: https://patents.google.com/patent/WO2014011042A1/en (accessed on 18 May 2022).
17. Wu, Y.; Li, W.; Martin, G.J.O.; Ashokkumar, M. Mechanism of low frequency and high-frequency ultrasound-induced inactivation of soy trypsin inhibitors. *Food Chem.* **2021**, *360*, 130057. [CrossRef]
18. Mulet-Cabero, A.-I.; Egger, L.; Portmann, R.; Ménard, O.; Marze, S.; Minekus, M.; Le Feunteun, S.; Sarkar, A.; Grundy, M.M.-L.; Carrière, F.; et al. A standardised semi-dynamic in vitro digestion method suitable for food—An international consensus. *Food Funct.* **2020**, *11*, 1702. [CrossRef]
19. Brodkorb, A.; Egger, L.; Alminger, M.; Alvito, P.; Assunção, R.; Ballance, S.; Bohn, T.; Bourlieu-Lacanal, C.; Boutrou, R.; Carrière, F.; et al. INFOGEST static in vitro simulation of gastrointestinal food digestion. *Nat. Protoc.* **2019**, *14*, 991–1014. [CrossRef]
20. Zaeim, D.; Mulet-Cabrero, A.; Read, S.A.; Liu, W.; Han, J.; Wilde, P.J. Effect of oil droplet size on the gastric digestion of milk protein emulsions using a semi-dynamic gastric model. *Food Hydrocoll.* **2022**, *124*, 107278. [CrossRef]
21. Minekus, M.; Alminger, M.; Alvito, P.; Ballance, S.; Bohn, T. A standardised static in vitro digestion method suitable for food- An international consensus. *Food Funct.* **2014**, *5*, 1113–1124. [CrossRef]
22. Fernández-García, E.; Lérida-Carvajal, I.; Pérez-Gálvez, A. In vitro bioaccessibility assessment as a prediction tool of nutritional efficiency. *Nutr. Res.* **2009**, *29*, 751–760. [CrossRef] [PubMed]
23. Rutherford, S.M. Methodology for determining degree of hydrolysis of proteins in hydrolysates: A review. *J. AOAC Int.* **2010**, *93*, 1515–1522. [CrossRef]
24. Andlinger, D.J.; Röscheisen, P.; Hengst, C.; Kulozik, U. Influence of pH, temperature and protease inhibitors on kinetics and mechanism of thermally induced aggregation of potato proteins. *Foods* **2021**, *10*, 796. [CrossRef] [PubMed]
25. Van Koningsveld, G.A.; Gruppen, H.; de Jongh, H.H.J.; Wijngaards, G.; van Boekel, M.A.J.S.; Walstra, P.; Voragen, A.G.J. Effects of Ph and heat treatments on the structure and solubility of potato proteins in different preparations. *J. Agric. Food. Chem* **2001**, *49*, 4889–4897. [CrossRef]
26. Tanger, C.; Andlinger, D.J.; Brümer-Rolf, A.B.; Engel, J.; Kulozik, U. Quantification of protein-protein interactions in highly denatured whey and potato protein gels. *MethodsX* **2021**, *8*, 101243. [CrossRef]
27. Schmidt, J.M.; Damgaard, H.; Greve-Poulsen, M.; Sunds, A.V.; Larsen, L.B.; Hammershøj, M. Gel properties of potato protein and the isolated fractions of patatin and protease inhibitors—Impact of drying method, protein concentration, pH and ionic strength. *Food Hydrocoll.* **2019**, *96*, 246–258. [CrossRef]
28. Elahi, R.; Mu, T.H. High hydrostatic pressure (HHP)-Induced structural modification of patatin and its antioxidant activities. *Molecules* **2017**, *22*, 438. [CrossRef]
29. Melnikov, S.M.; Stoyanov, S.D.; Kovacs, E.M.; Arnaudov, L.; De Groot, P.; Schuring, E.A.; Wiseman, S.A.; Mela, D.J.; Peters, H.P. Sustained hunger suppression from stable liquid foams. *Obesity* **2014**, *22*, 2131–2136. [CrossRef]
30. Ménard, O.; Famelart, M.H.; Deglaire, A.; Le Gouar, Y.; Guérin, S.; Malbert, C.H.; Dupont, D. Gastric emptying and dynamic in vitro digestion of drinkable yogurts: Effect of viscosity and composition. *Nutrients* **2018**, *10*, 1308. [CrossRef]
31. Shelat, K.J.; Nicholson, T.; Flanagan, B.M.; Zhang, D.; Williams, B.A.; Gidley, M.J. Rheology an microstructure characterization of small intestinal digesta from pigs fed a red meat-containing Western-style diet. *Food Hydrocoll.* **2015**, *45*, 300–308. [CrossRef]
32. Murray, K.; Placidi, E.; Schuring, E.A.H.; Hoad, C.; Koppenol, W.; Arnaudov, L.N.; Blom, W.; E Pritchard, S.; Stoyanov, S.D.; Gowland, P.A.; et al. Aerated drinks increase gastric volume and reduce appetite as assessed by MRI: A randomized, balanced, crossover trial. *Am. J. Clin. Nutr.* **2015**, *101*, 270–278. [CrossRef] [PubMed]
33. Lomolino, G.; Vincenzi, S.; Zannoni, S.; Marangon, M.; De Iseppi, A.; Curioni, A. Emulsifying activity of potato proteins in the presence of k-carrageenan at different pH conditions. *Food Chem. X* **2022**, *13*, 100232. [CrossRef] [PubMed]
34. Blanco, A.; Blanco, G. *Medical Biochemistry: Digestion-Absorption*; Academic Press: Cambridge, MA, USA, 2017; Chapter 12. [CrossRef]
35. Jiménez-Munoz, L.; Brodkorb, A.; Gómez-Mascaraque, L.; Corredig, M. Effect of heat treatment on the digestion behavior of pea and rice protein dispersions and their blends, studied using the semi-dynamic INFOGEST digestion method. *Food Funct.* **2021**, *12*, 8747–8759. [CrossRef] [PubMed]
36. Peram, M.R.; Loveday, S.M.; Ye, A.Q.; Singh, H. In vitro gastric digestion of heat-induced aggregates of beta-lactoglobulin. *J. Dairy Sci.* **2013**, *96*, 63–74. [CrossRef]

37. Singh, T.K.; Oiseth, S.K.; Lundin, L.; Day, L. Influence of heat and shear induced protein aggregation on the in vitro digestion rate of whey proteins. *Food Funct.* **2014**, *5*, 2686–2698. [CrossRef]
38. Zhang, Z.; Zhang, R.; McClements, D.J. Control of protein digestion under simulated gastrointestinal conditions using biopolymer microgels. *Food Res. Int.* **2017**, *100*, 86–94. [CrossRef]
39. Golding, M. Exploring and exploiting the role of food structure in digestion. In *Interdisciplinary Approaches to Food Digestion*; Gouseti, O., Bornhorst, G., Bakalis, S., Mackie, A., Eds.; Springer: Berlin/Heidelberg, Germany, 2019. [CrossRef]
40. Kim, M.Y.; Son, C.W.; Shim, H.J.; Lee, J.H.; Lee, K.J.; Sok, D.E.; Kim, H.C.; Yoon, W.K.; Kim, H.M.; Kim, M.R. Effect of cultivars and cooking methods on the trypsin inhibitor activities of potatoes. *Food Sci. Biotechnol.* **2008**, *17*, 161–165.
41. Avilés-Gaxiola, S.; Chuck-Hernández, C.; Serna-Saldivar, S.O. Inactivation methods of trypsin inhibitor in legumes: A review. *J. Food Sci.* **2017**, *83*, 17–29. [CrossRef]
42. Vanga, S.K.; Wang, J.; Raghavan, V. Effect of ultrasound and microwave processing on the structure, in-vitro digestibility and trypsin inhibitor activity of soymilk proteins. *LWT* **2020**, *131*, 109708. [CrossRef]
43. Nagase, H.; Salvesen, G.S. *Finding, Purification, and Characterization of Natural Protease Inhibitors in Proteolytic Enzymes: A Practical Approach*, 2nd ed.; Beynon, R.J., Bond, J.S., Eds.; Oxford University Press: Oxford, NY, USA, 2001; pp. 131–147.
44. Da Costa, H.P.S.; Oliveira, J.T.A.; Sousa, D.O.B.; Morais, J.K.S.; Moreno, F.B.; Monteiro-Moreira, A.C.O.; Viegas, R.A.; Vasconcelos, I.M. JcTI-I: A novel trypsin inhibitor from Jatropha curcas seed cake with potential for bacterial infection treatment. *Front. Microbiol.* **2014**, *5*, 5. [CrossRef]
45. Rivera del Rio, A.; Keppler, J.K.; Boom, R.M.; Janssen, A.E.M. Protein acidification and hydrolysis by pepsin ensure efficient trypsin-catalyzed hydrolysis. *Food Funct.* **2021**, *12*, 4570–4581. [CrossRef] [PubMed]
46. Chew, C.S. Parietal Cells. In *Encyclopedia of Gastroenterology*; Johnson, L.R., Ed.; Elsevier: Amsterdam, The Netherlands, 2004; pp. 140–144, ISBN 9780123868602. [CrossRef]
47. Rakhshi, E.; Nau, F.; Hiolle, M.; Floury, J. Pepsin diffusion in complex food matrices. *J. Food Eng.* **2022**, *324*, 111011. [CrossRef]
48. Pęksa, A.; Miedzianka, J. Potato industry by-products as a source of protein with beneficial nutritional, functional, health-promoting and antimicrobial properties. *Appl. Sci.* **2021**, *11*, 3497. [CrossRef]
49. Pęksa, A.; Kita, A.; Kułakowska, K.; Aniołowska, M.; Hamouz, K.; Nemś, A. The quality of protein of coloured fleshed potatoes. *Food Chem.* **2013**, *131*, 2960–2966. [CrossRef]
50. Shimomura, Y.; Murakami, T.; Nakai, N.; Nagasaki, M.; Harris, R.A. Exercise promotes BCAA catabolism: Effects of BCAA supplementation on skeletal muscle during exercise. *J. Nutr.* **2004**, *134*, 1583S–1587S. [CrossRef]
51. Holeček, M. Branched-chain amino acids in health and disease: Metabolism, alterations in blood plasma, and as supplements. *Nutr. Metab.* **2018**, *15*, 33. [CrossRef]
52. He, T.; Spelbrink, R.E.J.; Witteman, B.J.; Giussepin, M.L.F. Digestion kinetics of potato protein isolates in vitro and in vivo. *Food Sci. Nutr.* **2013**, *64*, 787–793. [CrossRef]
53. Herreman, L.; Nommensen, P.; Pennings, B.; Laus, M.C. Comprehensive overview of the quality of plant-and animal sourced proteins based on the digestible indispensable amino acid score. *Food Sci. Nutr.* **2020**, *8*, 5379–5391. [CrossRef]
54. Oikawa, S.Y.; Bahniwal, R.; Holloway, T.M.; Lim, C.; McLeod, J.C.; McGlory, C.; Baker, S.K.; Phillips, S.M. Potato protein isolate stimulates muscle protein synthesis at rest and with resistance exercise in young women. *Nutrients* **2020**, *125*, 1235. [CrossRef]

nutrients

MDPI

Article

In Vitro Digestibility of Minerals and B Group Vitamins from Different Brewers' Spent Grains

Anca Corina Fărcaş [1], Sonia Ancuţa Socaci [1], Maria Simona Chiş [1,*], Javier Martínez-Monzó [2], Purificación García-Segovia [2], Anca Becze [3], Anamaria Iulia Török [3], Oana Cadar [3], Teodora Emilia Coldea [1] and Marta Igual [2,*]

1 Department of Food Science, Faculty of Food Science and Technology, University of Agricultural Sciences and Veterinary Medicine of Cluj-Napoca, 400372 Cluj-Napoca, Romania

2 Food Investigation and Innovation Group, Food Technology Department, Universitat Politècnica de València, Camino de Vera s/n, 46022 Valencia, Spain

3 National Institute for Research and Development of Optoelectronics INOE 2000, Research Institute for Analytical Instrumentation, 67 Donath Street, 400293 Cluj-Napoca, Romania

* Correspondence: simona.chis@usamvcluj.ro (M.S.C.); marigra@upvnet.upv.es (M.I.); Tel.: +40-264-596384 (M.S.C.); +34-96-3879694 (M.I.)

Abstract: Brewers' spent grain (BSG), the main by-product of the brewing industry, is a rich source of minerals and water-soluble vitamins such as thiamine, pyridoxine, niacin, and cobalamin. Bioaccessibility through *in vitro* digestion is an important step toward the complete absorption of minerals and B group vitamins in the gastrointestinal system. Inductively coupled plasma optical emission spectrometry (ICP-OES) together with inductively coupled plasma quadrupole mass spectrometry (ICP-MS) was used for the quantification of the macro- and micro-minerals. An ultra-high performance liquid chromatography (UHPLC) system coupled with a diode array detector (DAD) was used for B group vitamin identification. Four different industrial BSG samples were used in the present study, with different percentages of malted cereals such as barley, wheat, and degermed corn. Calcium's bioaccessibility was higher in the BSG4 sample composed of 50% malted barley and 50% malted wheat (16.03%), while iron presented the highest bioaccessibility value in the BSG2 sample (30.03%) composed of 65% Pale Ale malt and 35% Vienna malt. On the other hand, vitamin B1 had the highest bioaccessibility value (72.45%) in the BSG3 sample, whilst B6 registered the lowest bioaccessibility value (16.47%) in the BSG2 sample. Therefore, measuring the bioaccessibilty of bioactive BSG compounds before their further use is crucial in assessing their bioavailability.

Keywords: *in vitro* digestibility; bioaccessibility; brewers' spent grain; minerals; B vitamins

1. Introduction

Food waste management represents a serious and ever-growing issue with important consequences for economic, environmental, and climatic changes. Even though progress has been made over time, there is an urgent demand for sustainability in the food processing chain [1]. To address this issue, the European Commission underlined the necessity of a Circular Economy Action Plan and mentioned that a sustainable proposal for circular economy might be the reuse of by-products as new raw materials [2].

Brewers' spent grain (BSG) is the main by-product obtained from the industrial brewing process, and EU brewing alone generates annually a total amount of 3.4 million tons of BSG [3]. According to Andres et al. [4], the production of one hectoliter of beer generates 20 kilos of BSG, and the number of breweries has recently doubled in the EU, nowadays reaching a total number of 9500. Worldwide, the beer industry has undergone great growth, reaching total production and consumption rates of 1.95 billion hectoliters and 357 million hectoliters, respectively [4].

It is worth pointing out that BSG represents 85% of the brewing waste and is a valuable source of bioactive compounds such as minerals, vitamins, fiber (hemicellulose, lignin, cellulose), proteins, and amino acids [3,5]. BSG contains important amounts of B group vitamins such as thiamine (B_1), riboflavin (B_2), niacin (B_3), pyridoxine (B_6), pantothenic (B_5), and folic acids, but also higher amounts of minerals such as calcium (Ca), sodium (Na), and magnesium (Mg), together with manganese (Mn), iron (Fe), potassium (K), and copper (Cu) [6]. A large body of studies has mentioned that BSG could be successfully used in the manufacturing of bakery and pastry products such as pasta, biscuits, cookies [7–9], bread [10], and extruded snacks [11], as well as meat products such as frankfurters [12] and burgers [13].

Even if the by-products are generally rich in bioactive compounds, their bioaccessibility is crucial to establish their nutritional value during the physiological digestion process [14]. Nowadays, more than two billion people all over the world are affected by different mineral deficiencies, with Fe and Zn deficiencies being the most prevalent for children [15]. Furthermore, Fe deficiency is considered the leading nutritional disorder worldwide according to Cian et al. [16], which affects over 30% of the world's population, mainly children and women from developing countries [17]. On the other hand, Zn deficiency is considered a worldwide problem, with an approximate percentage of risk for the populations of underdeveloped countries of 33% [17]. The consumption of food with bioaccessible minerals is essential for the human body in order to perform vital functions [18]. Even if there are several factors that could lead to mineral deficiencies such as an inadequate dietary intake, their poor bioavailability remains the most likely cause [19]. On the other hand, water-soluble vitamins such as B group are essential in preventing diseases such as anorexia, beriberi, mental problems, cardiovascular diseases, depression, visual disorders, complication of diabetes mellitus, or even dermatological disorders [20]. The bioavailability of the vitamins could not be fully predicted, mainly because of their unknown bioaccessibility during the gastrointestinal system [20].

Furthermore, bioaccessibility and bioavailability are the main concepts that must be taken into consideration to evaluate the pros and cons associated with the intake of an element from the food matrix and its further nutritional value in the organism [21]. The bioaccessibility is defined as the amount of bioactive compounds that are released from the food matrix before their absorption from the small intestine [21]. The bioavailability is mentioned as the fraction of ingested biocomponents that are able to reach the blood flow and the different organs and tissues, resulting in their bioactivity [22].

In a recent study, Igual et al. showed that an *in vitro* digestion process based on a static *in vitro* method such as COST INFOGEST network could be successfully used in evaluating the bioaccessibility of some bioactive compounds from the vegetable matrix [23].

The four BSG types were collected from Romanian industrial breweries, from different types of beers. The beers could be classified according to Salanță et al. [24] on three main groups, as follows: ale beers (top or high fermentation), lager with low fermentation, and non-alcoholic beers. On the other hand, Yeo et al. [25] mentioned different categories of craft beers or specialty beers, which include low-calorie beer, non-alcoholic and low-alcohol beers, gluten free beers, and functional and novel-flavored beers, whilst [26] mentioned that the key difference between beers is the fermentation method, and classified beers in "brown beer" and "white beer". White beer, which is also known as wheat beer, is defined by the World Cup Style Guidelines as a product made with at least 50% malted wheat, which is top fermented and with a typical aroma and flavor [26]. White beers produced by international breweries are gaining more and more attention, with one out of ten beers sold in Germany being a wheat beer [27]. Even if barley is the main cereal in brewing, maize and wheat have drawn mainstream attention, being used as beer adjuvants. For example, maize used at a percentage of 30% for beer manufacturing led to an 8% total reduction in production costs [28].

Therefore, the aim of the present paper is to establish the bioaccessibility of minerals and B group vitamins of different BSG types through *in vitro* gastrointestinal digestion model. The illustrative design of the present research is displayed in Figure 1, as follows:

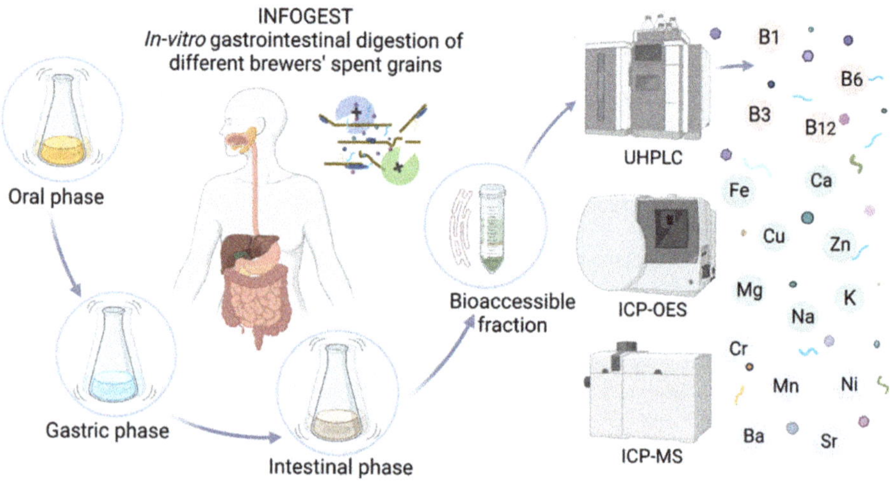

Figure 1. *In vitro* gastrointestinal digestion of minerals and B group vitamins from different BSG types.

2. Materials and Methods

2.1. Materials

The amylase, pepsin, gastric lipase, and pancreatin were purchased from Sigma-Aldrich Inc. (Sydney, NSW, Australia). All of the used solvents were of HPLC grade and were acquired from VWR, Inc. (Radnor, Pennsylvania, PA, USA). The ultra-pure water was obtained using an ULTRACLEAR UV UF EVOQUA purification system, (Erlanger, Kentucky, AL, USA). The multi-elemental solutions of 1000 mg/L ICP Standard Certipur® were purchased from Merk, (Darmstadt, Germany). All of the standards used were from Sigma-Aldrich (Sydney, NSW, Australia).

2.2. Brewers' Spent Grain from Different Types of Malt

The brewers' spent grain samples were kindly obtained from Romanian industrial breweries, which were dried for 12 h and milled using a professional laboratory mill (IKA A10, Staufen, Germany), as previously described by our research group [3]. The codification of the 4 by-products resulting from beer processing, the raw materials used, and the roasting treatments are displayed in Table 1.

Table 1. Types and codification of the selected brewers' spent grain samples.

Cereal By-Products	Used Grains	Percent	Kilning Temperature °C
BSG 1	Pale Ale Malt	40%	80–85
	Munich Malt	30%	100–105
	Caramel Malt	24%	220
	Black Malt	6%	235–250

Table 1. *Cont.*

Cereal By-Products	Used Grains	Percent	Kilning Temperature °C
BSG 2	Pale Ale Malt Vienna Malt	65% 35%	80–85 100–110
BSG 3	Pilsner Malt Degermed Corn	70% 30%	80–85 -
BSG 4	Pilsner Malt Wheat Malt	50% 50%	80–85 72–80

2.3. Simulated Gastrointestinal Digestion

The sample digestibility was assessed using the standardized static *in vitro* digestion method suitable for food (COST INFOGEST network) proposed by Minekus et al. [29] and Brodkorb et al. [30]. This *in vitro* method is a general, standardized, practical static digestion procedure based on relevant conditions that can be applied for various purposes. The aim of this procedure was to harmonize the *in vitro* static systems that simulate digestive processes by defining the key parameters and conditions.

The steps for the *in vitro* digestion protocol were: oral phase, mixing the sample and simulated salivary fluid (SSF) (1:1) with amylase (75 U/mL) at pH 7 for 2 min; gastric phase, mixing the oral bolus and simulate gastric fluid (SGF) (1:1) with pepsin (2000 U/mL) and gastric lipase (60 U/mL) at pH 3 for 2 h; intestinal phase, mixing the gastric chyme and simulate intestinal fluid (SIF) (1:1) with pancreatin (Trypsin activity 100 U/mL) at pH 7 for 2 h; filtration, centrifuging at 4500× *g* rpm for 30 min and then filtering through a 1 μm glass fiber membrane.

The concentration of enzymes used was estimated according to the activity certified by the manufacturer's analysis. The simulated fluids contained KCl, KH_2PO_4, $NaHCO_3$, NaCl, $MgCl_2$ $(H_2O)_6$, $(NH_4)_2CO_3$, $CaCl_2$, and HCL or NaOH for the pH adjustment of the stock solutions of simulated digestion fluids. The *in vitro* sample digestibility (IVD%) value was calculated as the difference between the initial sample and the undigested one; the obtained result was divided by the initial mass sample and multiplied by 100, as mentioned by [23,31].

The analyses were performed in triplicate. The blank and BSG samples were subjected to *in vitro* digestion, and the samples were collected according Minekus et al. [29] and Brodkorb et al. [30] and then freeze-dried using a protease inhibitor (Pefabloc SC, Sigma-Aldrich) with the ability to irreversibly inhibit trypsin and chymotrypsin. The blank and BSG digested samples were analyzed in triplicate for mineral and B group vitamins. The analyzed compounds present in the water and reagents were also analyzed and corrected in the final fraction. The bioaccessibility was calculated according to the proposed equation used by Khouzam et al. [32] as being the ratio between the concentration of bioactive compounds after *in vitro* digestion and the initial concentration of the same bioactive compounds before starting the *in vitro* digestion process.

2.4. Mineral Determination

In order to determine the major (Na, K, Ca, Mg, Fe, Cu, Zn) and trace element concentrations (Co, Cr, Mn, Ni, Ba, Sr, Rb), an Xpert closed-vessel system (Berghof, Eningen, Germany) was used for sample digestion. Here, 200 mg of sample was digested using 10 mL HNO_3 65% and 2 mL H_2O_2 30% in polytetrafluoroethylene digestion vessels, using a four-step digestion program (120 °C and 170 °C—heating; 100 °C and 25 °C—cooling) for a total digestion time of 25 min. After, the vessels were cooled down and the volume was made up with ultrapure water to 20 mL. The resulting solutions were analyzed using an Optima 5300DV inductively coupled plasma optical emission spectrometer (ICP-OES, Perkin Elmer, Norwalk, CT, USA) for the determination of the major elements, and an ELAN DRC II inductively coupled plasma quadrupole mass spectrometer (ICP-MS, Perkin–

Elmer, Waltham, MA, USA) for the determination of the minor elements. The calibration standards were prepared from ICP Multi-Element Standard Solution IV (1000 mg/L) (Merck, Darmstadt, Germany) for the ICP-OES and Multi-Element Calibration Standard 3 (10 mg/L) (Perkin Elmer Pure Plus, Waltham, MA, USA) at appropriate dilutions. The detection limits for the ICP-OES and ICP-MS are mentioned in the Supplementary Files, Table S1.

2.5. B Vitamins Analysis

For sample extraction, 0.5 g of sample was extracted with 1 mL of ultra-pure H_2O for 20 min in an ultrasonic bath (SONOREX, Bandelin, RK 103H, Darmstadt, Germany) at room temperature. The samples were centrifuged (Microcentrifuge Hettich D-78532, Germany) at 11,000× *g* rpm for 2 min, then the supernatant was filtered through a 0.45 µm cellulose filter.

A Vanquisher H UHPLC (ultra-high performance liquid chromatography) system from Dionex (Thermo Fisher Scientific, Germany) with a DAD detector was used for the analysis of the following vitamins: thiamine (B1), riboflavin (B2), nicotinamide (B3), pyridoxine (B6), cyanocobalamin (B12). The mobile phase was composed of ultra-pure H_2O with 1% acetic acid and MeOH in gradient with a flow of 0.3 mL/min. The chromatographic column used was a Accucore aQ (100 × 2.1 mm, 2.6 mm) from Thermo Fisher kept at 25 °C. The injection volume was 8 µL and the detector was set at 270 nm. Thiamine hydrochloride, riboflavin, niacinamide, pyridoxine hydrochloride, and cyanocobalamin standards were purchased form Sigma-Aldrich, Inc. (Sydney, NSW, Australia). The dilutions were made in ultra-pure water with a 0.1–100 µg/mL linearity range. The quantification limit was 0.1 and the detection limit was 0.03 µg/mL for all analyzed vitamins.

2.6. Statistical Analysis

An analysis of variance (ANOVA) with a 95% confidence level ($p < 0.05$) using the Statgraphics Centurion XVIII Software (Statgraphics Technologies, Inc., The Plains, VA, USA), version 18.1.13, was applied to evaluate the differences between the studied samples. The method used to discriminate between the means was Fisher's least significant difference procedure. All samples were made in triplicate and each result is expressed as the mean between them followed by the standard deviation. Small letters indicate homogeneous groups established by the ANOVA ($p < 0.05$) by comparing all samples (BSGs) before and after *in vitro* digestion, whilst capital letters indicate homogeneous groups established by the ANOVA ($p < 0.05$) by comparing samples before and after the *in vitro* digestion for each BSG sample.

3. Results

A total number of 18 minerals were determined, from which minerals such as cobalt (Co), arsenic (As), cadmium (Cd), mercury (Hg), lead (Pb), and rubidium (Rb) were below the detection limit (0.10 mg/kg).

The identified minerals were divided in two groups, as follows: macro-minerals such as calcium (Ca), potassium (K), sodium (Na), and magnesium (Mg); micro-minerals such as iron (Fe), zinc (Zn), manganese (Mn), copper (Cu), chromium (Cr), nickel (Ni), barium (Ba), and strontium (Sr). Macro-minerals are defined as minerals required in amounts greater than 100 mg/day, while micro-minerals, also entitled trace elements, are required in amounts of less than 100 mg/day [15].

With respect to the macro-minerals, Ca was the main mineral identified in BSG2 with a value of 2834.30 mg/kg (Figure 2), while Mg (Figure 2) was identified in bigger amounts in the BSG3 sample.

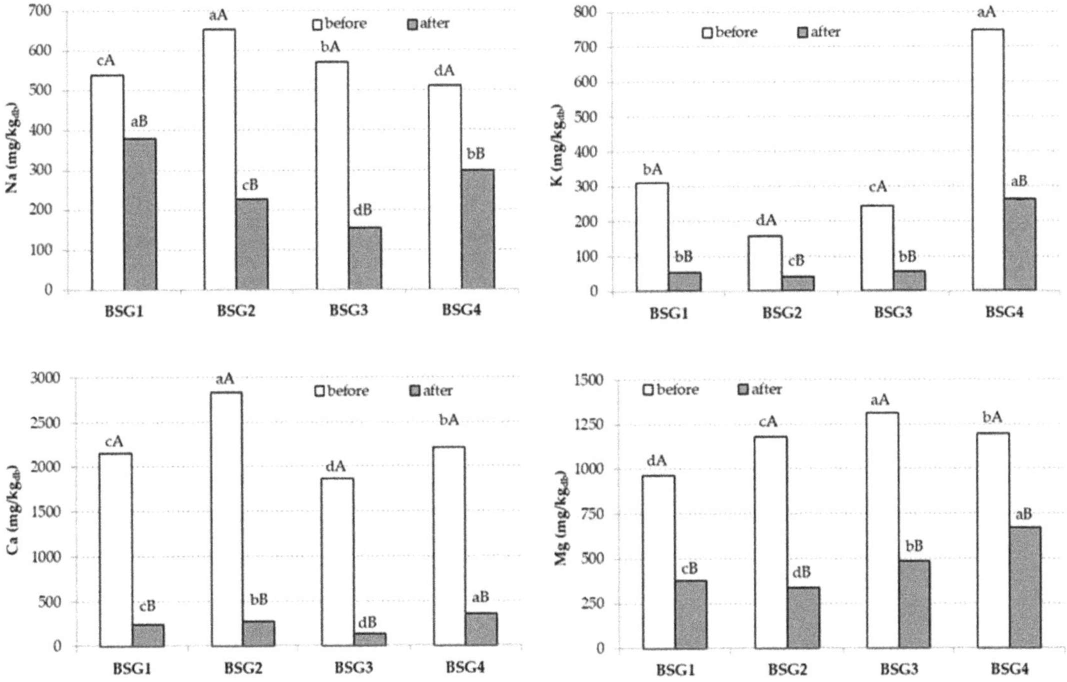

Figure 2. Mean values and standard deviations (error bars) of the macro-mineral (Na, K, Ca, and Mg) contents in four different brewers' spent grain (BSG) samples. For each mineral, small letters indicate homogeneous groups established by the ANOVA ($p < 0.05$) by comparing BSGs before and after *in vitro* digestion. For each mineral, capital letters indicate homogeneous groups established by the ANOVA ($p < 0.05$) by comparing samples before and after *in vitro* digestion for each BSG sample. Note: db, dry weight.

Regarding the Na and K contents in the studied BSG types, the values ranged between 509.93–652.82 mg/kg and 157.74–747.48 mg/kg, respectively. Similarly, Fărcaș et al. [3] showed that the Na and K contents could greatly vary between samples. For instance, Na range was 203.38–226.35 mg/kg and K was identified in a range of 486.14–514.37 mg/kg.

In the present study, Fe reached the highest value in the BSG1 sample (Figure 3), whilst Zn was the main representant of the BSG2 sample, with a value of 59.79 mg/kg, as illustrated in Figure 2. Mn reached higher extended values in BSG2 and BSG4 (Figure 2), whilst Cr was significantly higher in BSG4, with a value of 0.69 mg/kg (Figure 3). Sr reached similar values in BSG2 and BSG3 (Figure 3), while the Ba values were not statistically different between samples ($p < 0.05$) (Figure 3).

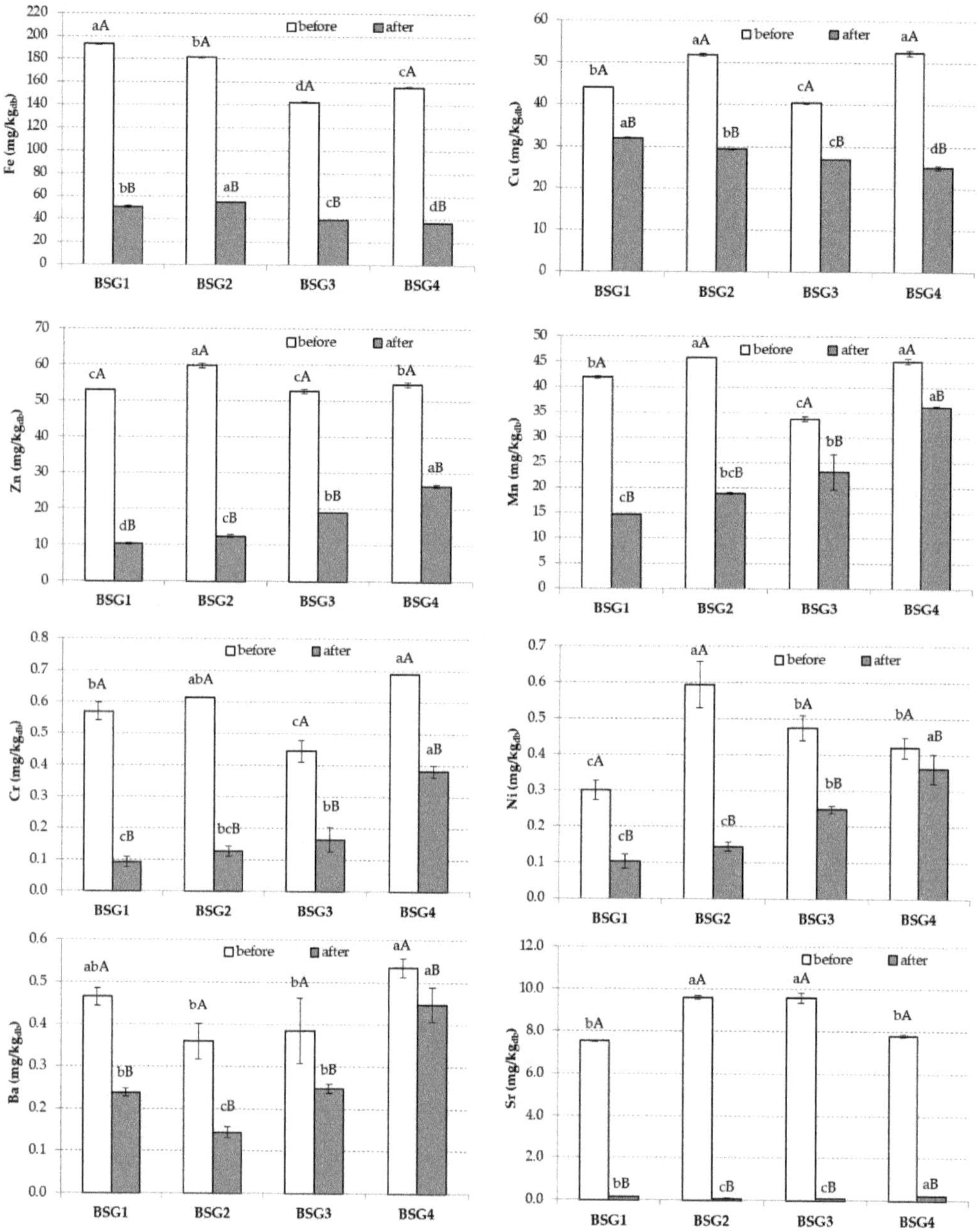

Figure 3. Mean values and standard deviations (error bars) of the micro-mineral (Fe, Cu, Zn, Mn, Cr, Ni, Ba, and Sr) contents in four different brewers' spent grain (BSG) samples. For each mineral, small letters indicate homogeneous groups established by the ANOVA ($p < 0.05$) by comparing BSGs before and after *in vitro* digestion. For each mineral, capital letters indicate homogeneous groups established by the ANOVA ($p < 0.05$) by comparing samples before and after *in vitro* digestion for each BSG sample. Note: db, dry weight.

For BSG4, bioaccessibility of Zn, Ca, and Mg reached values of 48.35%, 16.03%, and 56.07%, respectively, while the BSG2 sample registered lower values of 20.85%, 9.53%, and 28.44%, respectively. The bioaccessibility of Na was the highest in the BSG1 sample (70.43%), while a value of 58.43% was achieved in the BSG4 sample.

On the other hand, for BSG1, Cu bioaccessibility registered a statistically significant difference ($p < 0.05$) with a value of 72.42%, while BSG4 sample reached a value of 47.80%. With respect to Cr bioaccessibility, the highest value of 55.23% was obtained in the BSG4 sample, whilst a value of 16.05% was registered for BSG1 sample.

With respect to B group vitamins (B1, B3, B6, B12), their values before and after digestion are illustrated in Figure 4. The highest amount of vitamin B1 was reached in the BSG3 sample (Figure 4) with a value of 320.55 µg/g, whilst vitamins B6 and B12 were the highest in BSG1 and BSG2 samples, with values of 1057.36 µg/g and 550.93 µg/g, respectively (Figure 4). After the *in vitro* digestion, all analyzed vitamins decreased their values, as presented in Figure 3. Regarding their bioaccessibility, vitamin B1 had a value of 72.45% in the BSG3 sample, while a value of 68.44% was found for vitamin B3 for the BSG4 sample. Vitamin B6's bioaccessibility ranged from 16.47 to 52.94% for the BSG2 and BSG4 samples, respectively. Vitamin B12 registered values of 41.76%, 40.81%, 38.10%, and 83.57% for the BSG1, BSG2, BSG3, and BSG4 samples, respectively.

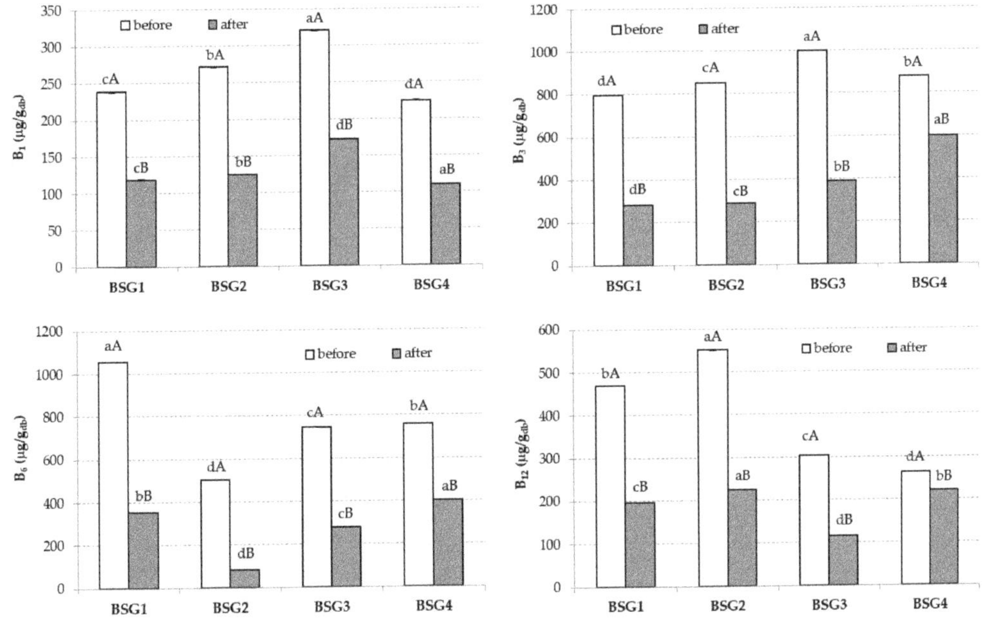

Figure 4. Mean values and standard deviations (error bars) of B group vitamin (B1, B3, B6, B12) contents in four different brewers' spent grain (BSG) samples. For each mineral, small letters indicate homogeneous groups established by the ANOVA ($p < 0.05$) by comparing BSGs before and after *in vitro* digestion. For each mineral, capital letters indicate homogeneous groups established by the ANOVA ($p < 0.05$) by comparing samples before and after *in vitro* digestion for each BSG sample. Note: db, dry weight.

4. Discussion

Bioaccessibility is defined as the rate-limiting step that influences the overall bioavailability of numerous bioactive compounds, as well as the portion released from the matrix and the amount available for further absorption [33]. The good reproducibility, lower cost, ease of sampling, labor and time savings, and lack of ethical restrictions for bioactive compounds are some of their advantages compared to in vivo studies [33].

The rich BSG mineral content was also highlighted by Jackowski et al. [34], who mentioned that the BSG ash amounts varied over a range of 1.1–4.6%, while Chetrariu and Dabija emphasized a value range of 2% to 6% for BSG ash [35]. Bonifácio-Lopes et al. highlighted that the Ca in BSG was identified within a range of 2200–3515 mg/kg, Mg at 1900–2400 mg/kg, and Na at 258.1–700 mg/kg [36]. In the present study, BSG Ca amounts ranged between 1857.4 and 2834.5 mg/kg, Mg was identified within a range of 963.79–1311.75 mg/kg, and Na at 537.11–569.66 mg/kg. It is worth pointing out that the BSG contents for almost all the macro-minerals and micro-minerals were statistically significantly different ($p < 0.05$) between samples (Figures 2 and 3). For instance, Ca, K, Na, and Mg for BSG3 of 1857.4 mg/kg, 242.64 mg/kg, 596.66 mg/kg, and 1311.75 mg/kg, respectively, while BSG4 had for the same macro-elements values of 2211.5 mg/kg, 748.48 mg/kg, 509.93 mg/kg, and 1196.08 mg/kg, respectively. Therefore, we can conclude that the tested BSG samples had variable mineral contents, leading to possible different influences on human health. The differences between the different BSG mineral contents could be explained by the processing method, or even the barley type used during malting, the harvesting and cultivation times, and the different adjuncts used during wort manufacturing [37,38].

Ca is a macromolecular nutrient that is necessary for the development, maintenance, and growth of human bone, and its deficiency leads to several diseases such as osteoporosis. Mg is an essential element that activates more than 300 enzymes involved in protein, nucleic acid, and carbohydrate metabolism, being essential for bone mineralization and development [39].

Fe is a micromineral that is indispensable for life, being involved in numerous physiological processes such as energy metabolism, oxygen transport, enzymatic activity, and even for the metabolic demands of renal cells [40]. Cr is considered an essential nutrient due to its presence in the active centers of several enzymes, being often used in the manufacture of different mineral or multivitamin supplements, with an RDA (recommended daily allowance) of 200 μg/day for adult women and men. Furthermore, Cr might be involved in the glucose transport from the blood to the cells, having positive effects on reducing blood sugar levels and being essential for complete glucose metabolism in humans [41,42].

On the other hand, Ni is an essential micro-nutrient for plants, having a main role in numerous biological mechanisms, and its deficiency can affect a plant's nitrogen metabolism and Fe absorption [43]. Ni also plays a key role in some enzymes, being an essential component [44]. Recently, Rosa et al. [45] mentioned that the WHO and FAO established a permissible limit for the Ni content in edible vegetable sources of 1.63 mg/kg, but our results are below the limit (0.30–0.60 mg/kg). Sr is recommended in some countries as a supplement for enhancing bone mineral density [46], whilst Zn is an essential mineral for the development and growth of all organisms, playing a main role in different cellular functions such as cell growth and division, cellular transport, and immune and endocrine systems [47].

As highlighted by Bohn et al. [48], there are several factors involved in mineral availability during digestion, such as oxidation and reduction or even complexation reactions. The bioaccessibility of minerals could be affected by their different chemical forms or by the presence of antinutritional substances that might interfere with their essential mineral absorption [18]. For instance, phytic acid, also entitled *myo*-inositol-hexaphosphate, is an antinutrient that was claimed to be present in brewers' spent grain and which is able to bind minerals and proteins, thereby influencing their solubility, digestibility, functionality, and absorption [49]. In line with this, Lynch et al. [50] mentioned that BSG has a high level of phytic acid, with a negative involvement in mineral solubility.

Moreover, it was claimed that *myo*-inositol phosphates could form insoluble complexes with divalent cations such as Ca^{2+}, Fe^{2+}, Mg^{2+}, Cu^{2+}, Mn^{2+}, and Zn^{2+}, having a highly negative influence on mineral absorption [18]. Furthermore, fiber could be another factor that might possibly lead to low mineral absorption [18], with BSG containing about 35–60%

fiber on a dry weight basis according to Puligundla et al. [51]. This idea is also sustained by Rousseau et al. [17], who emphasized fiber's capacity to bind minerals, forming fiber–mineral complexes, which are impossible to hydrolyze via human digestion. Fiber's specific characteristics, such as its solubility, might be the key factors that influence the mineral bioaccessibility level [17]. For instance, from the insoluble fiber types, lignin showed the greatest mineral-binding capacity as compared with hemicelluloses and cellulose, probably because of its polyphenolic nature [17]. In addition to fiber's capacity to bind minerals, it seems that its negative influence is also related to its capacity to influence the direct contact between nutrients and digestive enzymes, thanks to its increased digestive viscosity [52]. It should also be mentioned that BSG contains tannins [53], which could also be involved in mineral availability [18]. According to a recent study by Lemmens et al. [54], during human gastrointestinal digestion, phytate and arabinoxylan's cell wall materials resist, resulting in very low mineral bioaccessibility. Therefore, the initial chemical composition of the raw materials is one of the most important factors involved in their mineral bioaccessibility, mainly their antinutrients factors and their chelate digestibility (an antinutrient mineral complex) [17]. Moreover, it was highlighted that two main factors, i.e., the two different corn genotypes and the environmental conditions, could directly influence the phytate content in corn [55].

Cereal malting is also an important process, which involves germination followed by heat treatment, which generally lead to an increased nutritional content and degree of digestibility [19]. During malting, the mineral bioaccessibility could increase. For instance, the bioaccessibility of Ca and Mn in wheat (*Triticum aestivum*) increased through malting by 24% and 42%, while the bioaccessibility of Ca in barley remained the same during malting. This is in line with our result, where the biggest Ca bioaccessibility value was reached in the BSG4 sample (16.03%), composed of 50% malted barley and 50% malted wheat. With respect to Mn's bioaccessibility, the biggest value was reached in the BSG4 sample, followed by the BSG3 sample, with values of 80.08% and 68.81%, respectively.

It is of note that the bioaccessibility of Fe and Zn significantly increase during the germination process, mainly due to the phytate breakdown. For instance, wheat germination increased Fe and Zn's bioaccessibility from 4.6% and 2.5% to values of 14.1% and 14.6%, respectively [53]. Moreover, Lemmens et al. [54] showed that hydrothermal treatment could lead to an almost complete phytate breakdown, demonstrating that the Fe ions were no longer bound to the phytate structures after the treatment. In the present study, the highest Fe bioaccessibility was 30.03% for the BSG2 sample, whilst BSG4 presented the biggest value of 48.35% for Zn. This could also be justified by the heat treatment parameters, such as the malted kilning temperature, which could contribute to a reduction in antinutritional factors such as tannins, polyphenols, and phytic acid. Increased Fe bioaccessibility after heating was also mentioned by Raes et al. [56].

The absorption of Fe and Zn mainly happens in the small intestine, where Zn is absorbed by a carrier (ZnT-1)-mediated mechanism and non-heme Fe (identified in vegetable matrix) is transported across the mucosal membrane by divalent metal transporter 1 (DMT1), together with protons [17].

The low bioaccessibility of Ca (6.82%), Na (27.03%), and K (22.69%) for the BSG3 sample could be explained by the fact that corn was present at a percentage of 30% in the sample, and it was not previously malted. Even if corn is a rich source of minerals, it also contains some antinutritional factors such as trypsin, phytase inhibitors, and α-amylase, which might be involved in digestibility inhibition, absorption, and nutrient assimilation [28]. Additionally, for corn, there is a need for further processing in the beer industry, such as the removal of the oil-rich germ and bran, aiming to avoid poor foam formation [25].

BSG also contains water-soluble vitamins such as B1, B2, B3, B5, and B6 [52], which in the present study showed decreased values after the *in vitro* digestion. For instance, vitamin B1 in the BSG1 sample had an initial value of 238.52 µg/g d.w., and after the *in vitro* digestion its value decreased, registering a final amount of 117.85 µg/g d.w. (Figure 4). In line with this, vitamin B12 had the highest value in the BSG2 sample (550.9 μg/g d.w.), and after the *in vitro* digestion its value decreased more than twice, registering a final value of 224.83 μg/g d.w. (Figure 4). According to Yaman et al. [20], several factors could negatively affect the water-soluble B vitamins, such as the temperature, pH, dietary fiber, and grinding process, or even the polypeptides, polysaccharides, and antioxidants, but also the presence of metal ions and different digestive enzymes inhibitors.

Vitamin B1 is the most heat-responsive of all B group vitamins, and its bioaccessibility could decrease with the decreasing particle size of the dietary fibers. For instance, Kurek et al. [57] showed that vitamin B1′s bioaccessibility decreased from 69.1% to 91.2% with the decreasing dietary fiber particle size from 280 μm to 100 μm. Moreover, vitamins B3 and B6 bioaccessibility are also influenced by the particle size of the dietary fiber. In this idea, Akça et al. [58] recently showed that the bioaccessibility of vitamins B1, B2, and B3 is significantly influenced by the pH of the gastrointestinal tract, different polypeptide and polysaccharides bond, temperature, and stability. Moreover, kilning barley malt process is able to increase vitamins B1 and B2 according to Hassani et al. [59].

In the present study, the highest B1 bioaccessibility value was achieved in the BSG3 sample (72.45%), probably because wheat has high *in vitro* bioaccessibility in all parts of the seed, as previously mentioned by Zaupa et al. [60]. The authors mentioned *in vitro* bioaccessibility values for vitamin B1 ranging from 62.64% to 99.65%, with the highest value being quantified in the inner part of the aleurone wheat layer. On the other hand, the bioaccessibility of vitamin B3 in wheat ranged from 10.61% to 55.94%. In good agreement with this observation, our results showed that the bioaccessibility of vitamin B3 in the BSG3 sample was lower than for vitamin B1 (33.89%), probably because in the aleurone part, niacin forms inclusions with proteins, influencing their bioaccessibility [60].

Vitamin B6 could be present in food in several forms, including as pyridoxal, pyridoxine, pyridoxamine, or even pyridoxine-glucoside. The active form of vitamin B6 is pyridoxal 5-phosphate (PLP), which is considered a coenzyme for more than 100 metabolism and biochemistry enzymes [61]. Vitamin B6 is strictly correlated with proteins, having a non-covalently bond, and its separation is related to the pH values from the gastric and intestinal phases. Therefore, it can be stated that the bioaccessibility of PLP in adults is not as affected as that in infants, considering that adults have a lower pH value [61]. Our results showed that the biggest bioaccessibility value for vitamin B6 was 52.94% for the BSG4 sample, probably because wheat sprouts have greater amounts of B group vitamins than barley sprouts. Moreover, wheat sprouts contain pantothenic acid, which was not identified in barley sprouts [62].

Vitamin B12 is an essential water-soluble vitamin, being more resistant to heat treatment compared with other vitamins of the B group [20]. It has a leading role in cellular metabolism, acting as cofactor in several metabolic processes such as mitochondrial metabolism and DNA methylation and synthesis. Several diseases such as megaloblastic anemia or even neuropsychiatric ones could derive from vitamin B12 deficiency [63]. In the oral phase, vitamin B12 starts to be bound to haptocorrin, a specific protein from saliva and gastric juice, but in duodenum, due to the presence of pancreatic proteases, the complex is disrupted [64]. It could also be released in the gastric phase by hydrochloric acid and pepsin [20]. In the present study, the bioaccessibility of vitamin B12 resulted in similar values for BSG1 (41.79%) and BSG2 (40.81%), whilst BSG4 and BSG3 reached values of 83.57% and 38.10%, respectively. The lowest value was found for BSG3, which could be explained perhaps by the fact that the corn used in the beer manufacturing was not previously malted. Moreover, the amount of dietary fiber could be involved in decreasing the B group vitamins' bioaccessibility, since the hydroxyl group could react inside dietary fiber with the B group vitamins [65].

Therefore, the bioaccessibility of bioactive compounds such as minerals and vitamins is mainly related to the chemical composition of the food matrix. In this direction, several advances have been made over time, trying to increase the bioaccessibility and bioavailability of several food products before their in vivo or *in vitro* digestion. For instance, the use of solid-state fermentation in cereal or pulse by-products highly increased the phenolic bioaccessibility and bioavailability [66], whilst the germination increased the bioaccessibility of Ca, Fe, and Zn in pearl millet [67]. Additionally, Rousseau et al. [17] mentioned that milling, fermentation, dehulling, soaking, and thermal processing are other technological processes that could enhance the mineral bioaccessibility and bioavailability of plant-based foods, and Liao et al. [68] (Liao et al. 2022) showed that vitamin encapsulation might lead to a bioaccessibility improvement.

5. Conclusions

This study revealed that the contents of minerals and B group vitamins are affected by *in vitro* digestion. Micro-minerals such as Fe, Cu, Zn, Cr, Mn, Ni, Ba, and Sr and macro-minerals such as Ca, Mg, Na, K, and B group vitamins (B1, B3, B6, B12) presented different bioaccessibility values. From the macro-minerals, Na presented the biggest value (70.43%) for BSG1, whilst Cr presented the lowest bioaccessibility value (16.05%) for the same sample. Vitamin B12 showed the highest bioaccessibility (83.57%) in the BSG4 sample, followed by vitamin B1 with 72.45% bioaccessibility for the BSG3 sample and vitamin B3 with 68.44% bioaccessibility for the BSG3 sample and 52.94% bioaccessibility for the BSG4 sample.

To conclude, we can state that B12 and B1 have higher bioaccessibility levels compared with vitamins B3 and B6. The different BSG types and their chemical compositions (protein, fiber), together with the gastrointestinal pH system, temperature, and bound polypeptides and polysaccharides, are the main factors involved in the bioaccessibility of minerals and B vitamins.

Measuring bioactive compounds' bioaccessibility is a key factor in the manufacture of fortified foods containing different bioactive compounds, such as minerals and vitamins. New strategies need to be identified to increase the bioaccessibility of the bioactive compounds in brewers' spent grain before *in vitro* digestion. For instance, the use of modern extraction methods or the treatment of BSG with different enzymes before the digestion process might lead to a better improvement of the bioactive compounds' bioaccessibility. This research will shed light for further studies regarding the bioaccessibility of the minerals and B group vitamins in brewers' spent grains, with applications in several industries such as in food, pharmacology, or even cosmetics.

Supplementary Materials: The following supporting information can be downloaded at: https://www.mdpi.com/article/10.3390/nu14173512/s1, Table S1: Limits of detection for mineral determination by ICP–OES and ICP-MS.

Author Contributions: Conceptualization, A.C.F., S.A.S., M.S.C. and M.I.; methodology, A.C.F. and P.G.-S.; software, J.M.-M., A.B., A.I.T., M.I., O.C. and T.E.C.; validation, S.A.S., P.G.-S., A.B., A.I.T., O.C., T.E.C. and J.M.-M.; formal analysis, S.A.S., A.B., A.I.T., O.C. and M.I. investigation, A.C.F., S.A.S., J.M.-M., P.G.-S. and T.E.C.; resources, J.M.-M., A.C.F. and P.G.-S.; data curation, A.B., A.I.T., O.C. and T.E.C.; writing—original draft preparation, A.C.F. and M.S.C.; writing—review and editing, A.C.F. and M.I.; visualization, A.C.F., M.I. and S.A.S.; supervision, A.C.F.; project administration, A.C.F.; funding acquisition, A.C.F. and S.A.S. All authors have read and agreed to the published version of the manuscript.

Funding: This work was supported by a grant of the Romanian Ministry of Education and Research CCCDI—UEFISCDI, project number PN-III-P2-2.1-PED-2019-3622, within PNCDI III.

Institutional Review Board Statement: Not applicable.

Informed Consent Statement: Not applicable.

Data Availability Statement: Not applicable.

Conflicts of Interest: The authors declare no conflict of interest.

References

1. Tan, Y.X.; Mok, W.K.; Chen, W.N. *In Vitro* Evaluation of Enriched Brewers' Spent Grains Using Bacillus subtilis WX-17 as Potential Functional Food Ingredients. *Appl. Biochem. Biotechnol.* **2021**, *193*, 349–362. [CrossRef] [PubMed]
2. Allegretti, C.; Bellinetto, E.; D'arrigo, P.; Griffini, G.; Marzorati, S.; Rossato, L.A.M.; Ruffini, E.; Schiavi, L.; Serra, S.; Strini, A.; et al. Towards a Complete Exploitation of Brewers' Spent Grain from a Circular Economy Perspective. *Fermentation* **2022**, *8*, 151. [CrossRef]
3. Fărcaș, A.C.; Socaci, S.A.; Chiș, M.S.; Pop, O.L.; Fogarasi, M.; Păucean, A.; Igual, M.; Michiu, D. Reintegration of brewers spent grains in the food chain: Nutritional, functional and sensorial aspects. *Plants* **2021**, *10*, 2504. [CrossRef] [PubMed]
4. Andres, A.I.; Petron, M.J.; Lopez, A.M.; Timon, M.L. Optimization of extraction conditions to improve phenolic content and *in vitro* antioxidant activity in craft brewers' spent grain using response surface methodology (rsm). *Foods* **2020**, *9*, 1398. [CrossRef]
5. Amoriello, T.; Mellara, F.; Galli, V.; Amoriello, M.; Ciccoritti, R. Technological properties and consumer acceptability of bakery products enriched with brewers⇔ spent grains. *Foods* **2020**, *9*, 1492. [CrossRef]
6. Ikram, S.; Huang, L.Y.; Zhang, H.; Wang, J.; Yin, M. Composition and Nutrient Value Proposition of Brewers Spent Grain. *J. Food Sci.* **2017**, *82*, 2232–2242. [CrossRef]
7. Nocente, F.; Taddei, F.; Galassi, E.; Gazza, L. Upcycling of brewers' spent grain by production of dry pasta with higher nutritional potential. *Lwt* **2019**, *114*, 108421. [CrossRef]
8. Guo, M.; Du, J.; Zhang, Z.; Zhang, K.; Jin, Y. Optimization of Brewer's spent grain-enriched biscuits processing formula. *J. Food Process Eng.* **2014**, *37*, 122–130. [CrossRef]
9. Heredia-Sandoval, N.G.; Granados-Nevárez, M.d.C.; Calderón de la Barca, A.M.; Vásquez-Lara, F.; Malunga, L.N.; Apea-Bah, F.B.; Beta, T.; Islas-Rubio, A.R. Phenolic Acids, Antioxidant Capacity, and Estimated Glycemic Index of Cookies Added with Brewer's Spent Grain. *Plant Foods Hum. Nutr.* **2020**, *75*, 41–47. [CrossRef]
10. Fărcaş, A.C.; Socaci, S.A.; Tofană, M.; Mureşan, C.; Mudura, E.; Salanţă, L.; Scrob, S. Nutritional properties and volatile profile of brewer's spent grain supplemented bread. *Rom. Biotechnol. Lett.* **2014**, *19*, 9705–9714.
11. Ktenioudaki, A.; Crofton, E.; Scannell, A.G.M.; Hannon, J.A.; Kilcawley, K.N.; Gallagher, E. Sensory properties and aromatic composition of baked snacks containing brewer's spent grain. *J. Cereal Sci.* **2013**, *57*, 384–390. [CrossRef]
12. Özvural, E.B.; Vural, H.; Gökbulut, I.; Özboy-Özbaş, Ö. Utilization of brewer's spent grain in the production of Frankfurters. *Int. J. Food Sci. Technol.* **2009**, *44*, 1093–1099. [CrossRef]
13. Spinelli, S.; Conte, A.; Del Nobile, M.A. Microencapsulation of extracted bioactive compounds from brewer's spent grain to enrich fish-burgers. *Food Bioprod. Process.* **2016**, *100*, 450–456. [CrossRef]
14. Herrera-Cazares, L.A.; Luzardo-Ocampo, I.; Ramírez-Jiménez, A.K.; Gutiérrez-Uribe, J.A.; Campos-Vega, R.; Gaytán-Martínez, M. Influence of extrusion process on the release of phenolic compounds from mango (*Mangifera indica* L.) bagasse-added confections and evaluation of their bioaccessibility, intestinal permeability, and antioxidant capacity. *Food Res. Int.* **2021**, *148*, 110591. [CrossRef]
15. Affonfere, M.; Chadare, F.J.; Fassinou, F.T.K.; Linnemann, A.R.; Duodu, K.G. *In-vitro* Digestibility Methods and Factors Affecting Minerals Bioavailability: A Review. *Food Rev. Int.* **2021**, 1–29. [CrossRef]
16. Cian, R.E.; Proaño, J.L.; Salgado, P.R.; Mauri, A.N.; Drago, S.R. High iron bioaccessibility from co-microencapsulated iron/ascorbic acid using chelating polypeptides from brewers' spent grain protein as wall material. *Lwt* **2021**, *139*, 110579. [CrossRef]
17. Rousseau, S.; Kyomugasho, C.; Celus, M.; Hendrickx, M.E.G.; Grauwet, T. Barriers impairing mineral bioaccessibility and bioavailability in plant-based foods and the perspectives for food processing. *Crit. Rev. Food Sci. Nutr.* **2020**, *60*, 826–843. [CrossRef]
18. Silva, J.G.S.; Rebellato, A.P.; Caramês, E.T.d.S.; Greiner, R.; Pallone, J.A.L. *In vitro* digestion effect on mineral bioaccessibility and antioxidant bioactive compounds of plant-based beverages. *Food Res. Int.* **2020**, *130*, 108993. [CrossRef]
19. Platel, K.; Eipeson, S.W.; Srinivasan, K. Bioaccessible mineral content of malted finger millet (*Eleusine coracana*), wheat (*Triticum aestivum*), and barley (*Hordeum vulgare*). *J. Agric. Food Chem.* **2010**, *58*, 8100–8103. [CrossRef]
20. Yaman, M.; Çatak, J.; Uğur, H.; Gürbüz, M.; Belli, İ.; Tanyıldız, S.N.; Yıldırım, H.; Cengiz, S.; Yavuz, B.B.; Kişmiroğlu, C.; et al. The bioaccessibility of water-soluble vitamins: A review. *Trends Food Sci. Technol.* **2021**, *109*, 552–563. [CrossRef]
21. Delaqua, D.; Carnier, R.; Cadore, S.; Sanches, V.L.; Berton, R.S.; Corbi, F.C.A.; Coscione, A.R. *In vitro* bioaccessibility and bioavailability of selenium in agronomic biofortified wheat. *J. Food Compos. Anal.* **2022**, *105*, 104253. [CrossRef]
22. Dima, C.; Assadpour, E.; Dima, S.; Jafari, S.M. Bioavailability and bioaccessibility of food bioactive compounds; overview and assessment by *in vitro* methods. *Compr. Rev. Food Sci. Food Saf.* **2020**, *19*, 2862–2884. [CrossRef] [PubMed]
23. Igual, M.; Păucean, A.; Vodnar, D.C.; García-Segovia, P.; Martínez-Monzó, J.; Chiş, M.S. *In Vitro* Bioaccessibility of Bioactive Compounds from Rosehip-Enriched Corn Extrudates. *Molecules* **2022**, *27*, 1972. [CrossRef] [PubMed]
24. Salanță, L.C.; Coldea, T.E.; Ignat, M.V.; Pop, C.R.; Tofană, M.; Mudura, E.; Borșa, A.; Pasqualone, A.; Zhao, H. Non-Alcoholic and Craft Beer Production and Challenes. *Processes* **2020**, *8*, 1382. [CrossRef]
25. Yeo, H.Q.; Liu, S.Q. An overview of selected specialty beers: Developments, challenges and prospects. *Int. J. Food Sci. Technol.* **2014**, *49*, 1607–1618. [CrossRef]
26. Faltermaier, A.; Waters, D.; Becker, T.; Arendt, E.; Gastl, M. Common wheat (*Triticum aestivum* L.) and its use as a brewing cereal—A review. *J. Inst. Brew.* **2014**, *120*, 1–15. [CrossRef]

27. Mastanjević, K.; Šarkanj, B.; Krska, R.; Sulyok, M.; Warth, B.; Mastanjević, K.; Šantek, B.; Krstanović, V. From malt to wheat beer: A comprehensive multi-toxin screening, transfer assessment and its influence on basic fermentation parameters. *Food Chem.* **2018**, *254*, 115–121. [CrossRef]
28. Dabija, A.; Ciocan, M.E.; Chetrariu, A.; Codină, G.G. Buckwheat and Amaranth as Raw Materials for Brewing, a Review. *Plants* **2022**, *11*, 756. [CrossRef]
29. Minekus, M.; Alminger, M.; Alvito, P.; Ballance, S.; Bohn, T.; Bourlieu, C.; Carrière, F.; Boutrou, R.; Corredig, M.; Dupont, D.; et al. A standardised static *in vitro* digestion method suitable for food-an international consensus. *Food Funct.* **2014**, *5*, 1113–1124. [CrossRef]
30. Brodkorb, A.; Egger, L.; Alminger, M.; Alvito, P.; Assunção, R.; Ballance, S.; Bohn, T.; Bourlieu-Lacanal, C.; Boutrou, R.; Carrière, F.; et al. INFOGEST static *in vitro* simulation of gastrointestinal food digestion. *Nat. Protoc.* **2019**, *14*, 991–1014. [CrossRef]
31. Batista, A.P.; Niccolai, A.; Fradinho, P.; Fragoso, S.; Bursic, I.; Rodolfi, L.; Biondi, N.; Tredici, M.R.; Sousa, I.; Raymundo, A. Microalgae biomass as an alternative ingredient in cookies: Sensory, physical and chemical properties, antioxidant activity and *in vitro* digestibility. *Algal Res.* **2017**, *26*, 161–171. [CrossRef]
32. Khouzam, R.B.; Pohl, P.; Lobinski, R. Bioaccessibility of essential elements from white cheese, bread, fruit and vegetables. *Talanta* **2011**, *86*, 425–428. [CrossRef]
33. Wu, P.; Chen, X.D. Validation of *in vitro* bioaccessibility assays—A key aspect in the rational design of functional foods towards tailored bioavailability. *Curr. Opin. Food Sci.* **2021**, *39*, 160–170. [CrossRef]
34. Jackowski, M.; Niedzwiecki, L.; Trusek, A. Brewer' s Spent Grains—Valuable Beer Industry By-Product biomolecules Brewer' s Spent Grains—Valuable Beer Industry By-Product. *Biomolecules* **2020**, *10*, 1669. [CrossRef]
35. Chetrariu, A.; Dabija, A. Brewer's spent grains: Possibilities of valorization, a review. *Appl. Sci.* **2020**, *10*, 5619. [CrossRef]
36. Bonifácio-Lopes, T.; Teixeira, J.A.; Pintado, M. Current extraction techniques towards bioactive compounds from brewer's spent grain–A review. *Crit. Rev. Food Sci. Nutr.* **2020**, *60*, 2730–2741. [CrossRef]
37. Naibaho, J.; Korzeniowska, M. Brewers' spent grain in food systems: Processing and final products quality as a function of fiber modification treatment. *J. Food Sci.* **2021**, *86*, 1532–1551. [CrossRef]
38. Mussatto, S.I. Brewer's spent grain: A valuable feedstock for industrial applications. *J. Sci. Food Agric.* **2014**, *94*, 1264–1275. [CrossRef]
39. Ciosek, Ż.; Kot, K.; Kosik-Bogacka, D.; Łanocha-Arendarczyk, N.; Rotter, I. The effects of calcium, magnesium, phosphorus, fluoride, and lead on bone tissue. *Biomolecules* **2021**, *11*, 506. [CrossRef]
40. van Swelm, R.P.L.; Wetzels, J.F.M.; Swinkels, D.W. The multifaceted role of iron in renal health and disease. *Nat. Rev. Nephrol.* **2020**, *16*, 77–98. [CrossRef]
41. Pechova, A.; Pavlata, L. Chromium as an essential nutrient: A review. *Vet. Med.* **2007**, *52*, 1. [CrossRef]
42. Asli, M.; Azizzadeh, M.; Moghaddamjafari, A.; Mohsenzadeh, M. Copper, Iron, Manganese, Zinc, Cobalt, Arsenic, Cadmium, Chrome, and Lead Concentrations in Liver and Muscle in Iranian Camel (Camelus dromedarius). *Biol. Trace Elem. Res.* **2020**, *194*, 390–400. [CrossRef] [PubMed]
43. Kumar, A.; Jigyasu, D.K.; Kumar, A.; Subrahmanyam, G.; Mondal, R.; Shabnam, A.A.; Cabral-Pinto, M.M.S.; Malyan, S.K.; Chaturvedi, A.K.; Gupta, D.K.; et al. Nickel in terrestrial biota: Comprehensive review on contamination, toxicity, tolerance and its remediation approaches. *Chemosphere* **2021**, *275*, 129996. [CrossRef] [PubMed]
44. Mirabi, A.; Shokuhi Rad, A.; Nourani, S. Application of modified magnetic nanoparticles as a sorbent for preconcentration and determination of nickel ions in food and environmental water samples. *TrAC—Trends Anal. Chem.* **2015**, *74*, 146–151. [CrossRef]
45. Rosa, R.H.; Ribeiro Fernandes, M.; De Pádua Melo, E.S.; Granja Arakaki, D.; De Lima, N.V.; Santos Leite, L.C.; Espindola, P.R.; De Souza, I.D.; Do Nascimento, V.A.; Saldanha Tschinkel, P.F.; et al. Determination of Macro- And Microelements in the Inflorescences of Banana Tree Using ICP OES: Evaluation of the Daily Recommendations of Intake for Humans. *Sci. World J.* **2020**, *2020*, 8383612. [CrossRef]
46. Panahifar, A.; Chapman, L.D.; Weber, L.; Samadi, N.; Cooper, D.M.L. Biodistribution of strontium and barium in the developing and mature skeleton of rats. *J. Bone Miner. Metab.* **2019**, *37*, 385–398. [CrossRef]
47. Chasapis, C.T.; Ntoupa, P.S.A.; Spiliopoulou, C.A.; Stefanidou, M.E. Recent aspects of the effects of zinc on human health. *Arch. Toxicol.* **2020**, *94*, 1443–1460. [CrossRef]
48. Bohn, T.; Carriere, F.; Day, L.; Deglaire, A.; Egger, L.; Freitas, D.; Golding, M.; Le Feunteun, S.; Macierzanka, A.; Menard, O.; et al. Correlation between *in vitro* and in vivo data on food digestion. What can we predict with static *in vitro* digestion models? *Crit. Rev. Food Sci. Nutr.* **2018**, *58*, 2239–2261. [CrossRef]
49. Ainsworth, P.; Ibanoğlu, Ş.; Plunkett, A.; Ibanoğlu, E.; Stojceska, V. Effect of brewers spent grain addition and screw speed on the selected physical and nutritional properties of an extruded snack. *J. Food Eng.* **2007**, *81*, 702–709. [CrossRef]
50. Lynch, K.M.; Steffen, E.J.; Arendt, E.K. Brewers' spent grain: A review with an emphasis on food and health. *J. Inst. Brew.* **2016**, *122*, 553–568. [CrossRef]
51. Puligundla, P.; Mok, C. Recent advances in biotechnological valorization of brewers' spent grain. *Food Sci. Biotechnol.* **2021**, *30*, 341–353. [CrossRef]
52. Guillon, F.; Champ, M. Structural and physical properties of dietary fibres, and consequences of processing on human physiology. *Food Res. Int.* **2000**, *33*, 233–245. [CrossRef]

53. Puligundla, P.; Mok, C.; Park, S. Advances in the valorization of spent brewer's yeast. *Innov. Food Sci. Emerg. Technol.* **2020**, *62*, 102350. [CrossRef]
54. Lemmens, E.; De Brier, N.; Goos, P.; Smolders, E.; Delcour, J.A. Steeping and germination of wheat (*Triticum aestivum* L.). I. Unlocking the impact of phytate and cell wall hydrolysis on bio-accessibility of iron and zinc elements. *J. Cereal Sci.* **2019**, *90*, 102847. [CrossRef]
55. Baranwal, D. Malting: An indigenous technology used for improving the nutritional quality of grains— A review. *Asian J. Dairy Food Res.* **2017**, *36*, 179–183. [CrossRef]
56. Raes, K.; Knockaert, D.; Struijs, K.; Van Camp, J. Role of processing on bioaccessibility of minerals: Influence of localization of minerals and anti-nutritional factors in the plant. *Trends Food Sci. Technol.* **2014**, *37*, 32–41. [CrossRef]
57. Kurek, M.A.; Wyrwisz, J.; Karp, S.; Wierzbicka, A. Particle size of dietary fiber preparation affects the bioaccessibility of selected vitamin B in fortified wheat bread. *J. Cereal Sci.* **2017**, *77*, 166–171. [CrossRef]
58. Akça, S.N.; Sargın, H.S.; Mızrak, Ö.F.; Yaman, M. Determination and assessment of the bioaccessibility of vitamins B1, B2, and B3 in commercially available cereal-based baby foods. *Microchem. J.* **2019**, *150*, 104192. [CrossRef]
59. Hassani, A.; Procopio, S.; Becker, T. Influence of malting and lactic acid fermentation on functional bioactive components in cereal-based raw materials: A review paper. *Int. J. Food Sci. Technol.* **2016**, *51*, 14–22. [CrossRef]
60. Zaupa, M.; Scazzina, F.; Dall'Asta, M.; Calani, L.; Del Rio, D.; Bianchi, M.A.; Melegari, C.; De Albertis, P.; Tribuzio, G.; Pellegrini, N.; et al. *In vitro* bioaccessibility of phenolics and vitamins from durum wheat aleurone fractions. *J. Agric. Food Chem.* **2014**, *62*, 1543–1549. [CrossRef]
61. Yaman, M.; Mızrak, Ö.F. Determination and evaluation of *in vitro* bioaccessibility of the pyridoxal, pyridoxine, and pyridoxamine forms of vitamin B6 in cereal-based baby foods. *Food Chem.* **2019**, *298*, 125042. [CrossRef]
62. Eremina, O.Y.; Seregina, N.V.; Ivanova, T.N.; Shuldeshova, N.V.; Zaugolnikova, E.V. Micronutrient value and antioxidant activity of malt wheat sprouts. *IOP Conf. Ser. Earth Environ. Sci.* **2021**, *677*, 022107. [CrossRef]
63. Lavriša, Ž.; Hristov, H.; Hribar, M.; Žmitek, K.; Kušar, A.; Koroušić Seljak, B.; Gregorič, M.; Blaznik, U.; Gregorič, N.; Zaletel, K.; et al. Dietary Intake and Status of Vitamin B12 in Slovenian Population. *Nutrients* **2022**, *14*, 334. [CrossRef]
64. Etcheverry, P.; Grusak, M.A.; Fleige, L.E. Application of *in vitro* bioaccessibility and bioavailability methods for calcium, carotenoids, folate, iron, magnesium, polyphenols, zinc, and vitamins B 6, B 12, D, and E. *Front. Physiol.* **2012**, *3*, 317. [CrossRef]
65. Patil, R.N.; Nagaonkar, S.N.; Shah, N.B.; Bhat, T.S.; Almale, B. Study of perception and help seeking behaviour among parents for their children with psychiatric disorder: A community based cross-sectional study. *J. Med. Res.* **2016**, *2*, 6–11. [CrossRef]
66. Roasa, J.; De Villa, R.; Mine, Y.; Tsao, R. Phenolics of cereal, pulse and oilseed processing by-products and potential effects of solid-state fermentation on their bioaccessibility, bioavailability and health benefits: A review. *Trends Food Sci. Technol.* **2021**, *116*, 954–974. [CrossRef]
67. Luithui, Y.; Baghya Nisha, R.; Meera, M.S. Cereal by-products as an important functional ingredient: Effect of processing. *J. Food Sci. Technol.* **2019**, *56*, 1–11. [CrossRef]
68. Liao, P.; Dai, S.; Lian, Z.; Tong, X.; Yang, S.; Chen, Y.; Qi, W.; Peng, X.; Wang, H.; Jiang, L. The Layered Encapsulation of Vitamin B 2 and β -Carotene in Bioaccessibility of Vitamin B 2 and β -Carotene. *Foods* **2022**, *11*, 20. [CrossRef]

nutrients

MDPI

Article

Persimmon Proanthocyanidins with Different Degrees of Polymerization Possess Distinct Activities in Models of High Fat Diet Induced Obesity

Ying Yu [1,†], Ping Chen [1,†], Xiaofang Li [1], Shanshan Shen [2,*] and Kaikai Li [1,*]

[1] College of Food Science and Technology, Key Laboratory of Environment Correlative Food Science, Huazhong Agricultural University, Wuhan 430070, China
[2] Wuhan Children's Hospital, Tongji Medical College, Huazhong University of Science & Technology, Wuhan 430016, China
* Correspondence: author: shenshanshan1986@126.com (S.S.); likaikai@mail.hzau.edu.cn (K.L.)
† These authors contributed equally to this work.

Abstract: Proanthocyanidins is a kind of polyphenol that had been found with strong prevention ability on high fat diet induced obesity. However, whether proanthocyanidins with different polymerization degree showed different anti-obesity ability is unclear. Therefore, in this study, the effects of persimmon proanthocyanidins (P-PCs) and persimmon oligo-proanthocyanidins (P-OPCs) on high-fat diet induced obesity were systematically investigated. The findings indicated that both of P-PCs and P-OPCs significantly reduced the body weight, and P-PCs showed stronger anti-obesity ability compared with P-OPCs, P-OPCs seemed with stronger ability on improvement of insulin resistance. Furthermore, gut microbiota results indicated that the composition of the gut microbiota was changed after P-PCs and P-OPCs intervention in C57BL/6J mice. In addition, P-PCs exhibited strong inhibitory on the digestion of starch and fat. Above all, this study indicated that P-PCs showed stronger anti-obesity ability compared with P-OPCs.

Keywords: proanthocyanidins; polymerization degree; digestion; anti-obesity ability

1. Introduction

Obesity and its associated metabolic syndrome have become an important global public health problem [1–4]. Until now, developing effective innovative interventions is still an urgent issue. Application of some digestive enzyme inhibitors, which could prevent the absorption of nutrients, seems a useful strategy to alleviate obesity. Several naturally phytochemicals found in foods have been found with strong digestion enzyme inhibitory activities; for example polyphenols, which showed potential anti-obesity ability [5,6]. Proanthocyanidins (PCs) are kinds of polyphenols existing in many plant foods, and flavan-3-ols is the main structural unit which with various degrees of polymerization (DP) [7,8]. PCs have been found with many biological activities such as antioxidative, anti-obesity, antibacterial potential [6,8,9]. It has been found that PCs could inhibit the digestion enzyme, reshape intestinal microenvironment and prevent adipocyte differentiation, which played key roles for its anti-obesity potential.

Depending on the DP, PCs are classified as polymeric proanthocyanidins (P-PCs, DP > 5) and oligomeric proanthocyanidins (OPCs, DP 2 to 5). Previous results have found that PCs with DP > 4 could not be absorbed [10]. More importantly, it has been proven that the biological activities and bioavailability of PCs was directly influenced by the degrees of polymerization. For example, Pierini et al. found that PCs with different DP exhibited different anti-cancer ability in various cell lines [11]. Quiñones et al. found that OPCs with anti-hypertensive effect through adjusting the endothelium-derived NO bioavailability [12]. However, there were also other results, which indicated P-PCs with strong activities.

Bitzer et al. found that P-PCs from cocoa showed the strongest effective on gut barrier function and epithelial inflammation [13]. Gonçalves et al. also found that with the increase of the degree of polymerization, the inhibitory potential of α-amylase was increased; a high degree of polymerization made it more easy to form a stable interaction with a-amylase [14]. These findings highlighted the fact that the relationship between degrees of polymerization of PCs and their bio-activities were highly context dependent. However, until now, how the degrees of polymerization influence the anti-obesity of PCs is still unclear.

Persimmon is a widely consumed fruit with a high content of PCs [15]. A large number of studies have demonstrated persimmon PCs with various bio-activities, for example enzyme inhibitory ability, antioxidant activity, anti-obesity, and anti-inflammatory activity [15,16]. However, persimmon PCs were a kind of PC with high degree of polymerization (mDP = 26), which was insoluble in water and with a strong astringent taste; in the meanwhile, it was easy to connect with protein and polysaccharide, which seriously limited the usages of P-PCs [17]. In our previous study, we prepared P-OPCs from P-PCs with catalytic hydrogenation; the average DP of P-OPCs was 2.8 [16]. P-OPCs showed stronger solubility, antibacterial activities and free radical scavenging potential compared with P-PCs [16,18]. Thus, it is of interest to determine whether P-PCs and P-OPCs demonstrate different anti-obesity activities. Therefore, in this study, the potential effects of P-PCs and P-OPCs on high-fat diet induced obesity were investigated.

2. Materials and Methods

2.1. Catalytic Hydrogenolysis of P-PCs

Ripe persimmon fruit "GongchengYueshi" (*Diospyros kaki* L.) were obtained from a local market (Baishazhou Market, Wuhan city, China). P-PCs were extracted from whole fruit as previous described [15]. The preparation of P-OPCs was performed with catalytic hydrogenolysis method [16]. The hydrogenolysis condition was at 3 MPa, 120 °C, 3 h with 0.4 mg/mL of Pd/C; the average DP of P-PCs and obtained P-OPCs were 26 and 2.8 [16]. The detail information about the composition of oligomers of P-OPCs was as our previous study described [16].

2.2. In Vivo Animal Study

2.2.1. Animals and Dosage Regimen

All experiments were confirmed by the Laboratory Animal Research Center and the Ethics Committee of Huazhong Agricultural University. Male C57BL/6J mice (18–20 g) were kept in a stable environment (20 ± 2 °C, 12-h light/dark cycles) with free access to water and food. D12450B and D12492 were purchased from Readydietech. Co., Ltd. (Shenzhen, China). The detail information of the formula of D12450J and D12492 could be found in Supplementary Table S1. C57BL/6J mice were divided into four groups (n = 10): (1) Control group (C); (2) the High Fat Diet (HFD) group (M); (3) the HFD gavaged with 40 mg kg^{-1} of P-PCs group (PPCs); and (4) the HFD gavaged with 40 mg kg^{-1} of P-OPCs group (OPCs). Food intake and body weights were recorded twice a week.

2.2.2. Glucose Tolerance Tests (GTTs)

The Oral GTTs were investigated during week 9 in all four groups. Mice were fasted overnight and then gavaged with 2 g glucose/kg BW of glucose. At glucose administration, tail blood samples were collected at 0, 30, 60, 90, and 120 min, and the blood glucose content was measured with a blood glucose meter (Roche Diagnostics, Germany), as in our previous study.

2.2.3. Tissue Sample Collection

Animals were fasted overnight and anesthetized after the treatment. An eyeball blood sample was first collected, and then centrifuged at 3000 rpm for 10 min to prepare serum samples, then frozen at −80 °C. Livers and fat pad tissue were collected and weighed, then the tissue samples were cut into small pieces for fixed in 4% paraformaldehyde or stored at −80 °C for further analysis.

2.2.4. Biochemical Analyses and Histology

Serum triglycerides (TGs), total cholesterol (TC), low density lipoprotein (LDL), and high density lipoprotein (HDL) levels were investigated using enzymatic colorimetric methods with special reagent kits (Nanjing Jiancheng, China) according to the manufacturers' instructions. Fixed liver and adipose tissue were embedded in paraffin, then cut into sections with a thickness of 4 μm, and then stained with hematoxylin and eosin (H&E). Morphological changes were observed under a microscope.

2.3. DNA Extraction and Information Analysis of Gut Microbiota

After the mice were sacrificed, feces were quickly collected from the cecum and frozen in liquid nitrogen, then stored at −80 °C. Total microbial DNA was extracted using an E. Z.N.A™ Mag-Bind Soli DNA Kit (Omega Bio-tek, Norcross, GA, USA). The extracted DNA sample was accurately quantified using the Qubit3.0 DNA detection kit. The V3-V4 hypervariable region of the bacterial 16S rRNA gene were amplified with primer pairs (338F and 806R). Purified amplicons were paired-end sequence with an Illumina NovaSeq PE250 platform (Illumina, San Diego, CA, USA) by Majorbio Bio-Pharm Technology Co., Ltd. (Shanghai, China).

2.4. Enzyme Activity Assay

Assays of the inhibition of P-PCs and P-OPCs on activities of α-amylase and pancreatic lipase α-Amylase (Sigma-Aldrich, St. Louis, MO, USA) activity was investigated with dinitrosalicylic acid color reagent, pancreatic lipase (Sigma-Aldrich, St. Louis, MO, USA) activity was estimated using 4-MUO method as our previous report [5]. The 50% inhibition concentrations (IC_{50}) of P-PCs and P-OPCs were evaluated.

Fluorescence Quenching Measurements

The influences of PCs and P-OPCs on fluorescence spectra of α-Amylase and pancreatic lipase were investigated with Hitachi F-4600 fluorescence spectrometer as our previous described [19]. α-Amylase and pancreatic lipase were both prepared in PBS at 1 mg/mL. α-Amylase or pancreatic lipase with or without P-PCs and P-OPCs (0, 10, 20, 30, and 40 mg/L) was hatched at 37 °C for 15 min. Fluorescence emission were recorded in a 1 cm quartz-cell at λex of 278 nm, and λem of 290 to 500 nm for α-Amylase, and λex of 280 nm, and λem of 300 to 500 nm for Pancreatic lipase. Both excitation and emission slit widths were set at 5 nm and maintained with a scanning rate of 1200 nm/min. The Stern–Volmer equation was applied to investigate the fluorescence quenching mechanism.

2.5. Statistical Analysis

The results were presented as Mean ± SDs. All data were analyzed with one-way ANOVA using GraphPad Prism ver. 5.1 software (La Jolla, CA, USA). Statistical difference was set at $p < 0.05$.

3. Results

3.1. P-PCs Administration Showed Stronger Potential on HFD Induced Weight Gain Compared with P-OPCs

As the results showed in Figure 1a, the food intake of the mice in C groups was higher than that of the M group; the PPCs group was significantly lower than the M group (Figure 1a). The mice in HFD group showed a higher body weight than that in C group (42.17 ± 4.65 vs. 29.38 ± 2.86 g). The groups that administered with both of P-PCs and P-OPCs exhibited significant inhibitory potential in body weight (33.05 ± 3.02 vs. 42.17 ± 4.65 g and 36.65 ± 4.03 vs. 42.17 ± 4.65 g for PPCs and POPCs, respectively) (Figure 1b,c). In the meanwhile, HFD treatment could induce the increase of serum TC, TG and LDL. P-PCs and P-OPCs treatment significantly reduced the levels of serum TGs (0.90 ± 0.16 vs. 1.27 ± 0.21 mmol L^{-1}, $p < 0.01$) and TC (2.58 ± 0.57 vs. 4.16 ± 0.35 mmol/L, $p < 0.01$; 3.32 ± 0.38 vs. 4.16 ± 0.35 mmol/L, $p < 0.05$ for PPCs and POPCs, respectively) (Figure 1d,e), and P-PCs treatment group showed lower TC and Cho content compared with that of P-OPCs ($p < 0.05$). For LDL-C and HDL, P-PCs treatment could significantly reverse the change that induced by M (Figure 1f,g). For P-OPCs, even there was no significant different compared with the M treatment, P-OPCs treatment also exhibited improvement potential.

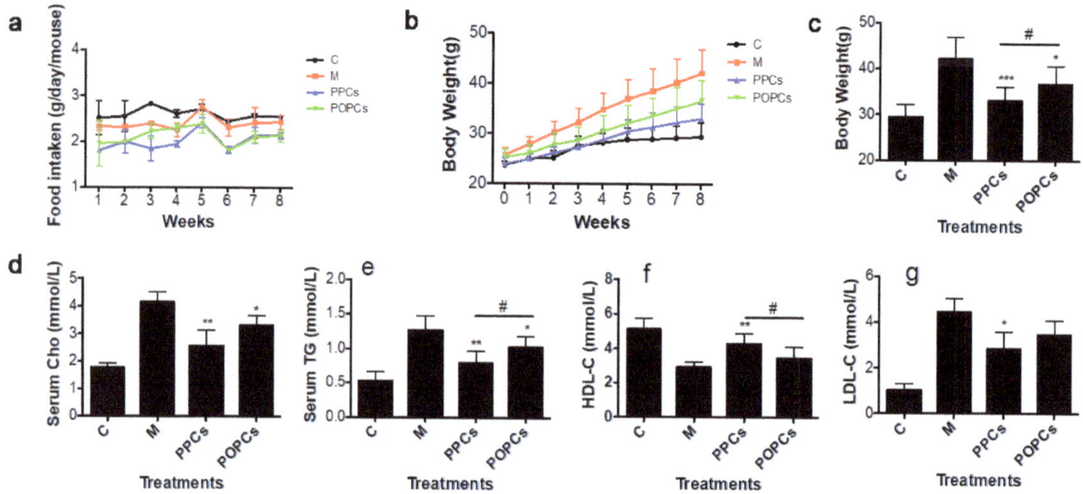

Figure 1. P-PCs and P-OPCs administration protect mice against excessive weight gain. C57BL/6 mice were fed a HFD with the administration of P-PCs and P-OPCs. (**a**) Food intake of mice with different treatment; (**b**) Body weight gain curve; (**c**) body weight at the last week; (**d–g**) Serum Cho, TG, HDL and LDL levels in different treatment groups. The results were analyzed with a one-way ANOVA with Tukey's Multiple Comparison Test using GraphPad Prism ver. 5.1 software (San Diego, CA, USA). * $p < 0.05$, ** $p < 0.01$, *** $p < 0.001$; # $p < 0.05$ between PPCs and POPCs.

3.2. P-PCs and P-OPCs Administration Improved Insulin Resistance

As shown in Figure 2a, the mice in M group showed higher fasting glucose levels (4.76 ± 0.53 vs. 8.17 ± 1.02 mmol/L, $p < 0.01$), and treatment with P-PCs and P-OPCs showed a significant decrease (5.99 ± 0.63 vs. 8.17 ± 1.02 mmol/L; 5.87 ± 0.52 vs. 8.17 ± 1.02 mmol/L, for P-PCs and P-OPCs, respectively). Compared with that in C group, the blood glucose levels were much higher in the M group after glucose administration (Figure 2b,c). The area under the curve (AUC) of glucose also indicated that both of P-PCs and P-OPCs could improve insulin resistance (AUC: 34.86 ± 1.24 vs. 43.48 ± 2.23 and 32.70 ± 1.82 vs. 43.48 ± 2.23 for P-PCs and P-OPCs, respectively).

3.3. Dietary P-PCs and P-OPCs Reduce Lipid Accumulation in HFD Induced Obese Mice

As the results showed in Figure 3a, both P-PCs and P-OPCs could significantly reduce the size and number of lipid droplets observed in the livers and epididymal fat mass. HFD could induce the increase of the weight of inguinal fat mass, epididymal fat mass, and perirenal fat mass, and both P-PCs and P-OPCs could inhibit the expansion of adipose tissue, compared with P-OPCs; P-PCs showed stronger inhibitory potential (Figure 3b–d). In addition, as shown in Figure 3b, the epididymal fat mass weights of mice in PPCs group were significantly lower than that in P-OPCs group (Figure 3b). Both P-PCs and P-OPCs treatment could decrease the cell area of epididymal fat mass induced by HFD, and compared with P-OPCs, P-PCs should stronger inhibitory potential (Figure 3e). In the meanwhile, as shown in Figure 3f,g, P-PCs and P-OPCs significantly decreased the liver weight and inhibited the accumulation of liver lipids compared with M group.

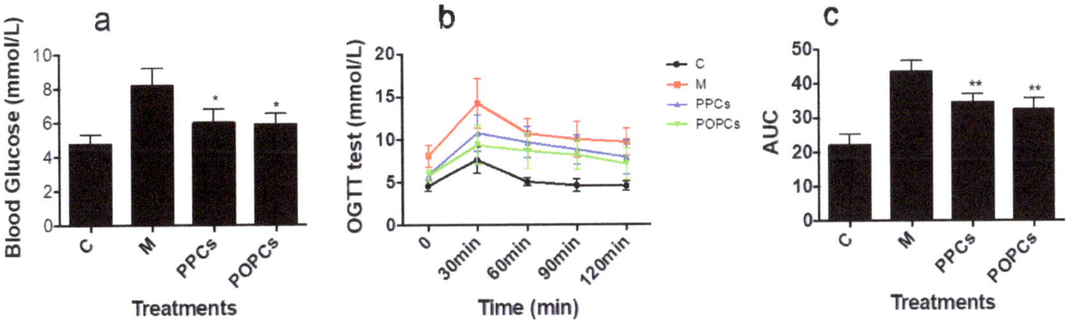

Figure 2. P-PCs and P-OPCs administration improve HFD induced insulin resistance. C57BL/6 mice were fed a HFD with the administration of P-PCs and P-OPCs. (**a**) Fasting blood glucose content; (**b**,**c**) GTT results after 8 weeks of P-PCs and P-OPCs administration. Blood glucose levels were measured at 0, 15, 30, 60, 90, and 120 min after oral administration of glucose (2 g kg^{-1}). The results were analyzed with a one-way ANOVA with Tukey's Multiple Comparison Test using GraphPad Prism ver. 5.1 software. * $p < 0.05$, ** $p < 0.01$.

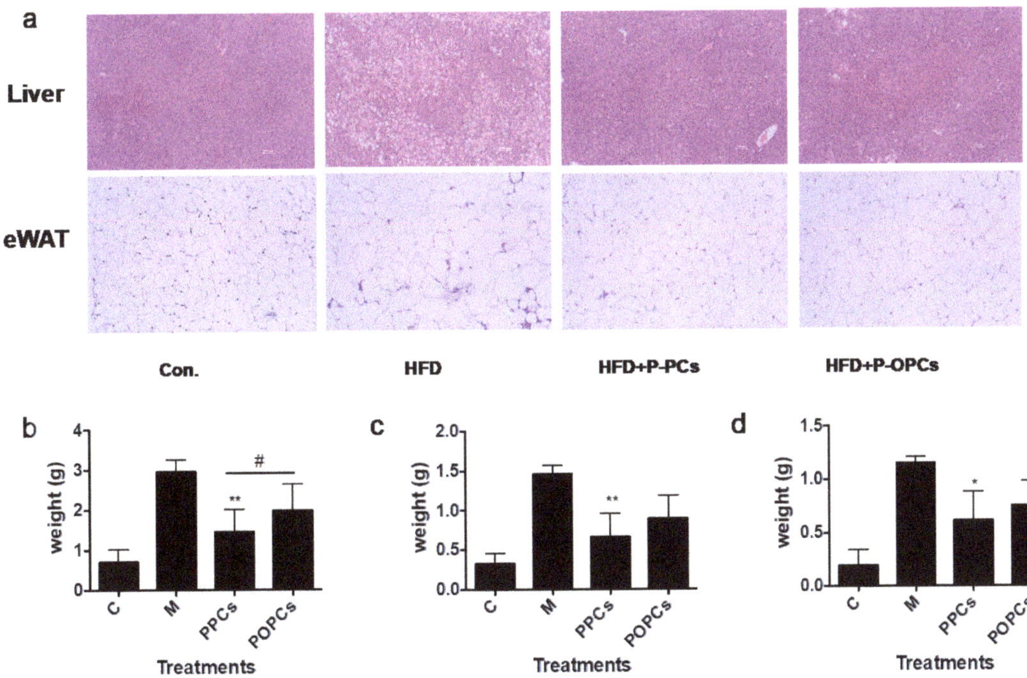

Figure 3. *Cont.*

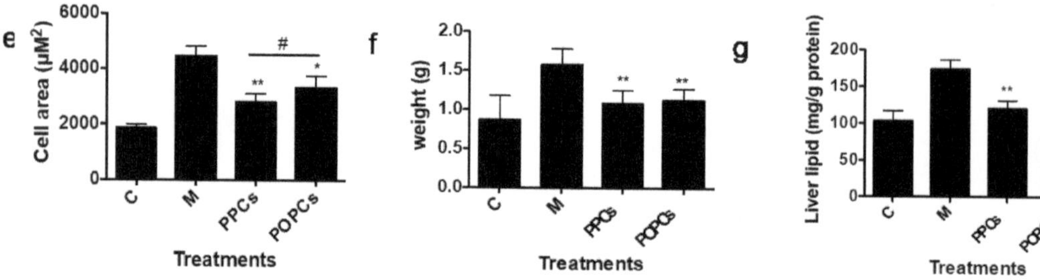

Figure 3. P-PCs and P-OPCs administration improves hepatic lipid accumulation and adipose tissue development. C57BL/6 mice were fed a HFD with the administration of P-PCs and P-OPCs. (**a**) H&E stained liver sections and epididymal adipose tissue; (**b–d**) weight of perirenal fat mass, epididymal fat mass and inguinal fat mass in different treatment groups; (**e**) Adipocyte size area of epididymal fat mass; (**f,g**) Liver weights and liver lipids accumulation in different group. Hepatic lipids were extracted by homogenizing the liver tissue in chloroform/methanol lipid extraction buffer and were analyzed according to the manufacturer's protocol. The results were analyzed with a one-way ANOVA with Tukey's Multiple Comparison Test using GraphPad Prism ver. 5.1 software. * $p < 0.05$, ** $p < 0.01$. # $p < 0.05$ between PPCs and POPCs.

3.4. Effect of P-PCs and P-OPCs on Gut Microbiota Diversity

Next, we further explore the impact P-PCs on the gut microbiota composition via 16S rRNA gene sequencing. There were 447, 402, 402, and 427 OTUs in the C group, M group, PPCs, and POPCs groups, respectively (Figure 4a). Accounting for 75% of the total, OTUs (313) were common existing among the four groups. For alpha diversity analysis, P-PCs treatment groups showed higher Shannon index and lower Simpson index compared with that in HFD group, which indicated that P-PCs treatment improved the alpha diversity decreased induced by M (Figure 4b–d). PCoA analysis of the four groups indicated that gut microbiota compositions were independent. The difference in the first (15.53%) and second (38.99%) principal component of gut microbiota indicated the difference in gut microbiota composition. Bray–Curtis assay showed the gut microbiota from C group, gut microbiota of M mice clustered separately with significant difference (Adonis, $p = 0.001$) (Figure 4e,f). The results showed that HFD intervention significantly changed the composition of the gut microbiota, however, there was no significant difference between P-PCs treatment groups and M group.

3.5. Composition of the Gut Microbiota

Phylum distributions of different treatment group were shown in Figure 5a, *Bacteroidetes* and *Firmicutes* were the primary bacteria (about +80%). The relative abundance of *Bacteroidetes* was significantly reduced in the M group (5.42 ± 0.39% vs. 17.46 ± 4.50%, $p < 0.05$), and *Firmicutes* was increased (79.38% ± 11.81% vs. 66.18 ± 5.88%, $p < 0.05$) compared with that in the C group. Compared with that in M group, P-PCs treatment could significantly increase the relative abundance of *Bacteroidetes* and *Epsilonbacteraeota* ($p < 0.05$). In P-OPCs treatment group, there was no significant difference compared with M; Bacteroidetes was also increased and *Proteobacteria* and *Actinobacteria* also decreased. There was no significant different between P-PCs and P-OPCs treatment groups (Figure 5b,c).

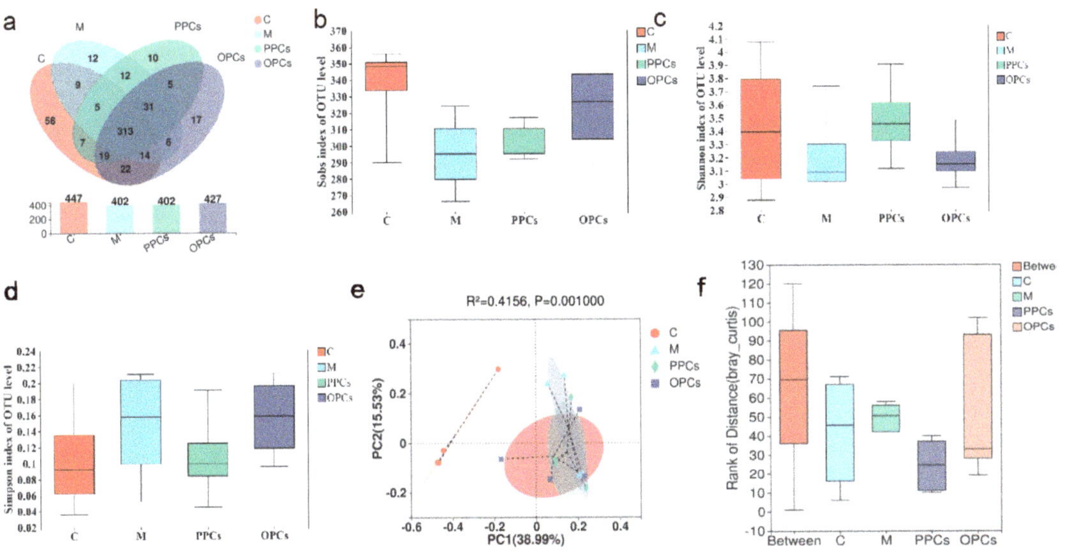

Figure 4. Gut microbiota composition of P-PCs and P-OPCs treatment mice. C57BL/6 mice were fed a HFD with the administration of P-PCs and P-OPCs. (**a**) OTU numbers of gut microbiota in different treatment groups; (**b**) Sobs indexes of OTU level; (**c**) Shannon indexes of OTU level; (**d**) Simpson indexes of OTU level; (**e**,**f**) PCoA analysis of the gut microbiota composition at OTU level.

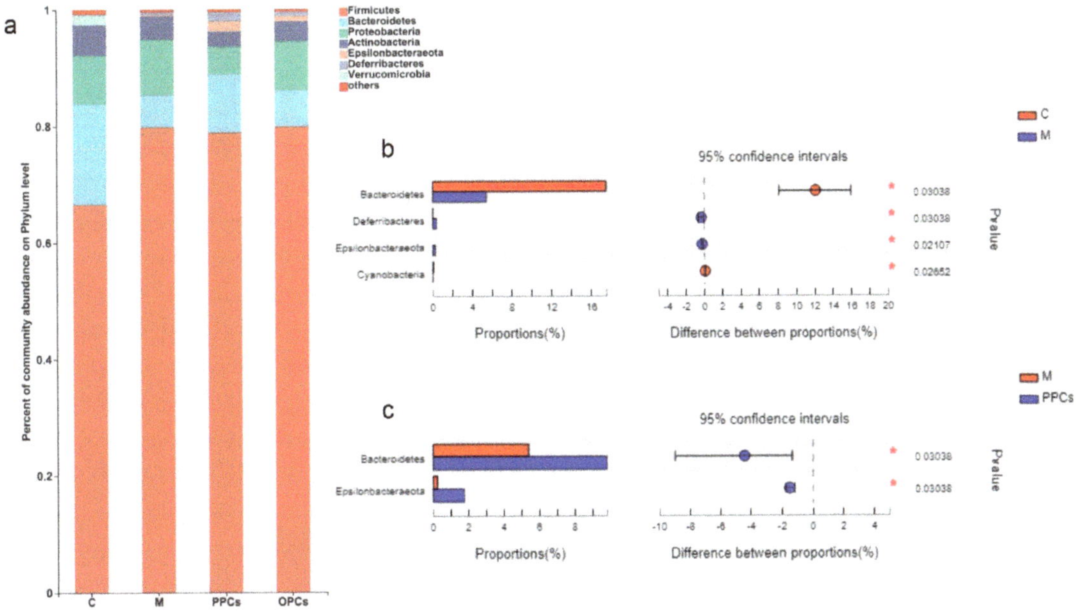

Figure 5. Gut microbiota composition at phylum level. (**a**) Relative abundance of gut microbiota at phylum level; (**b**) The differences of gut microbiota between C and M, (**c**) The differences of gut microbiota between M and PPCs. Data were analyzed by Wilcoxon rank-sum test, * $p < 0.05$.

3.5.1. Genus Level

As the results presented in Figure 6a–e show, the relative abundances of *Faecalibaculum*, *norank-f-Muribaculaceae*, and *unclassified f-Ruminococcaceae* showed significant reduction in the M group, whereas those of *Blautia* and *Lactobacillus* ($p < 0.05$) were significantly increased. In the PPCs and POPCs groups, the relative levels of *Helicobacter* and *Odoribacter* were significantly increased in the P-PCs group, while those of *Romboutsia* and *Ruminococcaceae* were significantly reduced compared with those in the HFD group. In the P-OPCs treatment group, the relative level of *Tyzzerella* was significantly increased, while those of *Romboutsia* was significantly reduced. Compared with the P-OPCs treatment group, the abundances of *Lactobacillus* was higher, and *unclassified_f_Family_XIII*, *Ruminococcaceae_UCG_009* and *Clostridium_sensu_stricto_1* were lower in the P-PCs treatment groups.

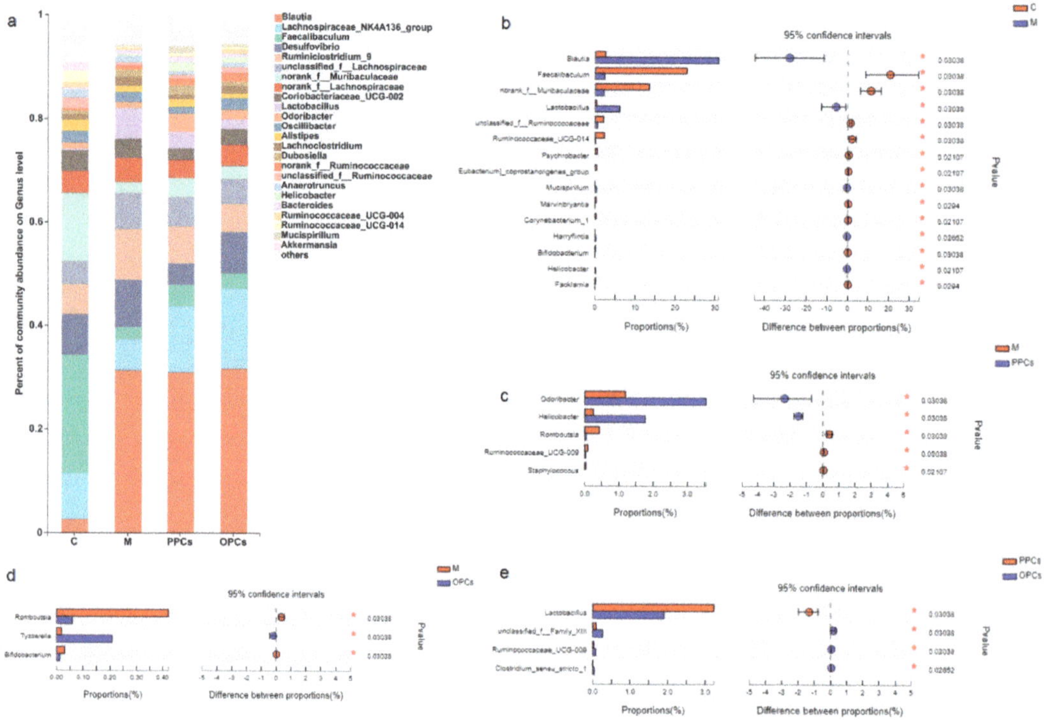

Figure 6. Gut microbiota composition at genus level. (**a**) Relative abundance of gut microbiota at genus level; (**b**) the differences of gut microbiota between C and M groups; (**c**) the differences of gut microbiota between M and PPCs groups; (**d**) the differences of gut microbiota between different M and POPCs groups; (**e**) the differences of gut microbiota between different PPCs and POPCs groups. Data were analyzed by Wilcoxon rank-sum test, * $p < 0.05$.

3.5.2. Species Level

As the results presented in Figure 7a–e show, the relative abundances of *uncultured_bacterium_g_Faecalibaculum*, *unclassified_f_Ruminococcaceae*, and *uncultured_bacterium_g_norank_f_Muribaculaceae* were significantly reduced in the M group. The relative abundance of *unclassified_g_Blautia* and *unclassified_g_Lactobacillus* were significantly increased in the mice of M group. Compared with M group, the relative abundances of unclassified_g_Odoribacter, Helicobacter_hepaticus, and unclassified_g_Lachnospiraceae were significantly increased; and uncultured_bacterium_g_Romboutsia, uncultured_bacterium_g Ruminococcaceae_UGC-009 were reduced in P-PCs treatment group. The P-OPCs treatment significantly increased the levels of uncultured_bacterium_g_Roseburia and uncultured_bacterium_g_Tyzzerella, and the relative abundances of uncultured_bacterium_g_Romboutsia was significantly lower compared with that in the HFD group. There were also many differences between P-PCs and P-OPCs treatment groups; for example, unclassified_g__Lactobacillus. These findings suggested that HFD could destroy the structure of the gut microbiota, P-PCs and P-OPCs could improve the gut microbiota disorder induced by HFD.

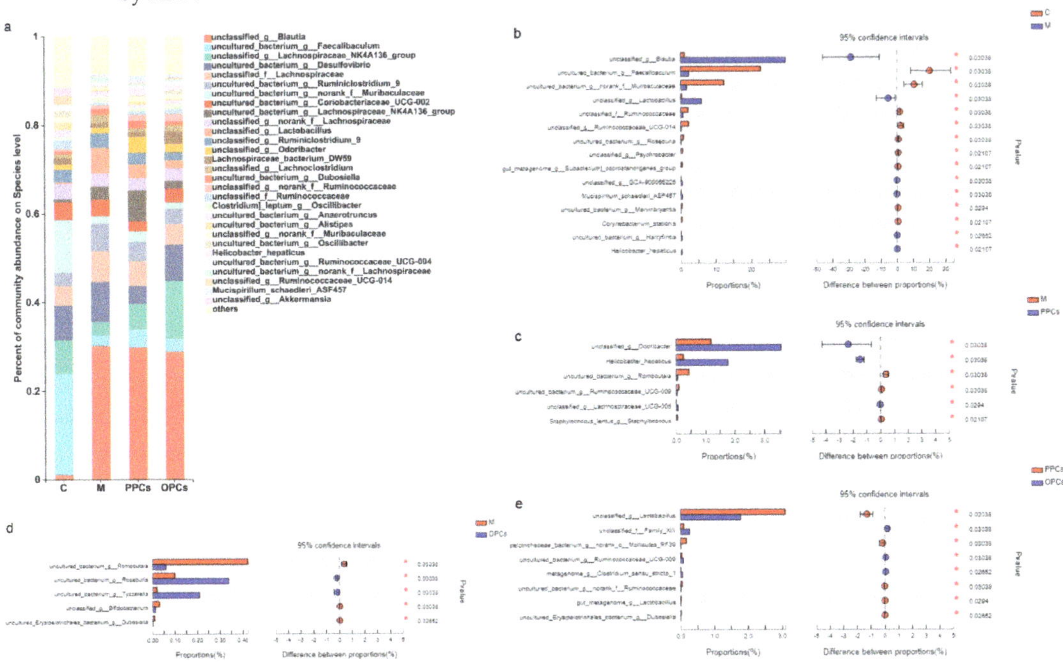

Figure 7. Gut microbiota composition at species level. (**a**) Relative abundance of gut microbiota at species level; (**b**) the differences of gut microbiota between C and M groups; (**c**) the differences of gut microbiota between M and PPCs groups; (**d**) the differences of gut microbiota between M and POPCs groups; (**e**) the differences of gut microbiota between PPCs and OPCs groups. Data were analyzed by Wilcoxon rank-sum test, * $p < 0.05$.

3.6. Inhibitory Activities of P-PCs and P-OPCs on Digestive Enzymes Activities

The inhibitory potentials of P-PCs and P-OPCs on α-amylase and pancreatic lipase were investigated. Compared with P-OPCs, P-PCs showed stronger inhibitory potential on pancreatic lipase and α-amylase; with the IC50 values were 78.6 μg/mL (P-PCs) and 105.4 μg/mL (P-OPCs) for α-amylase, 22.9 μg/mL (P-PCs) and 43.1 μg/mL (P-OPCs) for pancreatic lipase. The significant inhibitory ability of P-PCs and P-OPCs on α-amylase and pancreatic lipase suggested that P-PCs and P-OPCs might directly bind to the enzymes. Therefore, the binding characters between P-PCs and P-OPCs and α-amylase or pancreatic lipase were further investigated.

The fluorescence quenching spectra of α-amylase by P-PCs and P-OPCs at 37 °C were shown in Figure 8. A sturdy fluorescence emanation peak at 345 nm was found in α-amylase after exciting at $\lambda = 280$ nm. For α-amylase, the fluorescence strength of α-amylase markedly diminished with the increase of the dose of P-PCs and P-OPCs; the EC_{50} values were 10.94 mg/L of P-PCs and 16.57 mg/L of P-OPCs at 37 °C. P-PCs and P-OPCs also dose-dependently quenched the fluorescence of pancreatic lipase, and the EC_{50} was 17.85 mg/L of P-PCs and 21.58 mg/L of P-OPCs at 37 °C. Compared with P-OPCs, P-PCs showed large Ksv and Kq values both with α-amylase and pancreatic lipase, which indicated a stronger interaction between P-PCs and the enzymes molecular. These results indicated that P-PCs showed stronger quenching potential on α-amylase and pancreatic lipase compared with P-OPCs. These results were also consisted with previous studies that the fluorescence quenching capacity of grape seed proanthocyanidins on α-amylase were positively collected to its DP [14].

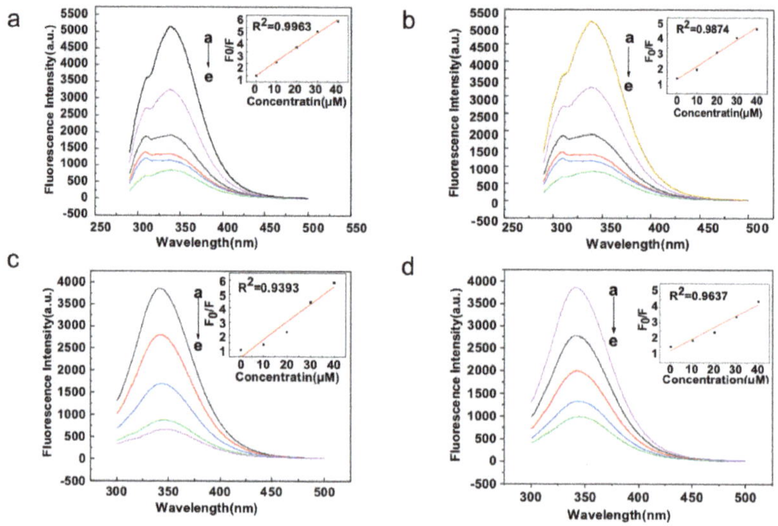

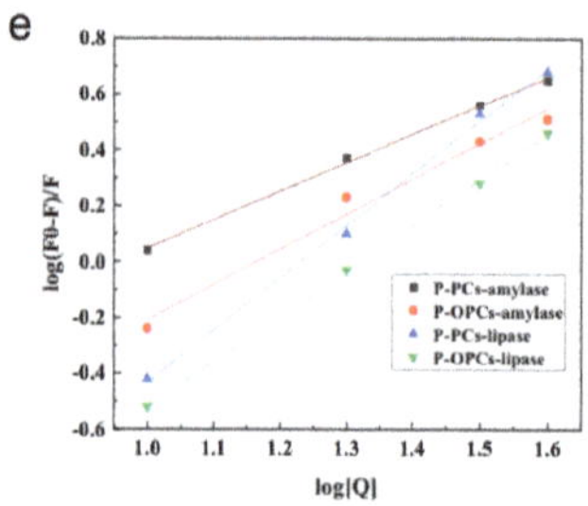

f

\	P-PCs-amylase	P-OPCs-amylase	P-PCs-lipase	P-OPCs-lipase
R^2_{Ksv}	0.9963	0.9874	0.9393	0.9637
Ksv (L mol^{-1})	1.16×10^5	0.86×10^5	1.28×10^5	0.73×10^5
Kq (Lmol^{-1}s^{-1})	1.16×10^{13}	0.86×10^{13}	1.28×10^{13}	0.73×10^{13}
R^2_a	0.9971	0.9729	0.9963	0.9938
Ka (Lmol^{-1})	1×10^5	0.16×10^5	0.05×10^5	0.04×10^5
n	1.04	1.56	1.9	1.85

Figure 8. Fluorescence quenching spectra of Key Digestive enzymes at 37°C with P-PCs and P-OPCs. (**a**,**b**) Fluorescence quenching spectra of α-amylase at 37 °C with P-PCs and P-OPCs; (**c**,**d**) Fluorescence quenching spectra of lipase at 37 °C with P-PCs (**c**) and P-OPCs (**d**); (**e**,**f**) quenching constant Kq-values of binding between α-amylase and lipase at 37 °C with P-PCs and P-OPCs.

4. Discussion

Proanthocyanidins exist in many food materials, for example, grape seed, persimmon, sorghum, and cocoa, mainly existing in the form of polymeric proanthocyanidins (>90%) [20,21]. In our previous study, the P-OPCs from P-PCs were prepared by the catalytic hydrogenation method, which demonstrates good solubility, bio-availability, and free radical scavenging ability. However, it has been found that the degrees of polymerization of proanthocyanidins played central role for the biological activities of PCs [6,14]. More importantly, the correlation between polymerization degree and bioactivity appears to be dependent on the different kinds of bioactivity. Therefore, in this study, the effects of P-PCs and P-OPCs on high-fat diet induced obesity were systematically investigated.

Nowadays, the rapidly increasing number of the population with obesity has become a serious health problem. In this study, a HFD induced obesity model was applied to systemically investigate the inhibitory properties of P-PCs and P-OPCs on diet induced obesity and the underlying mechanisms. The results indicated that both of P-PCs and

P-OPCs reduced the body weight, TC, TG, and LDL, while increasing the content of HDL in HFD treatment mice. Compared with P-OPCs, P-PCs showed stronger anti-obesity potential, which showed lower body weight, TG, TC, and so on. However, in this study, there was an interesting phenomenon regarding the obesity-related insulin resistance; treatment with P-PCs and P-OPCs decreased the glucose levels, even there was no significant difference, and the P-OPCs treatment group showed lower fasting glucose level; meanwhile, GTT test also showed similar results. These results indicated that P-PCs and P-OPCs could improve insulin resistance, and P-OPCs seemed with stronger ability compared with P-PCs.

Many studies consistently supported dietary PCs with health benefits in murine models and humans; for example, anti-obesity, improve insulin sensitivity. It had been found that the improvement of PCs on insulin sensitivity were not linked to inhibitory potential in obesity [6]. Paquette et al. found that a mix rich in PACs and phenolic acids extract could increase insulin sensitivity and lower first-phase insulin secretion; however, this treatment showed no improvement effects on body weight, lipid profile, and oxidative stress [22]. Stull et al. also found the similar results [23]. Hokayem et al. found 2 g of grape polyphenols per day could prevent fructose induced insulin resistance [24]. These findings were consisted with our results, even P-OPCs with lower ability in inhibition of body weight, P-OPCs seems with stronger ability for insulin sensitivity compared with P-PCs treatment. These results also indicated that the mechanism of improvement of insulin sensitivity of P-PCs and P-OPCs were different.

Recently, growing evidence has indicated that diet induced obesity was correlated with the gut microbiota. High fat diet could induce gut microbiota disorder by reducing the diversity and destroying the microbiota structure. In this study, the Shannon index and the Simpson index indicated that the alpha diversity increased in the P-PCs and P-OPCs treatment groups, though not significantly different. PCoA analysis also found that composition of the gut microbiota was changed in HFD group; however, there was no significant change between PCs treatment groups and HFD group. It has been reported that HFD could alter cecum microflora populations and increase the abundance of *Firmicutes*, and decrease the abundance of *Bacteroidetes*. These results were similar in this study. Compared with HFD group, P-PCs and P-OPCs treatments could increase the abundance of Bacteroidetes. Compared with HFD group, both of P-PCs and P-OPCs treatments could decrease the abundance of Romboutsia. Wang et al. found that HFD could significantly induced the increase of Romboutsia, and walnut green husk extract treatment could significantly inhibit the obesity along with the decrease of Romboutsia, this result was consisted with our finding [25]. Though there were many significant differences between P-PCs and HFD, P-OPCs and HFD were found; however, there were only a few gut microbiota shown in the same manner in the P-PCs and P-OPCs treatments group compared with HFD group. Therefore, we hypothesized the anti-obesity ability of P-PCs and P-OPCs were not always dependent on the blooming or decreasing of any bacterium, despite being influenced by P-PCs and P-OPCs; the change in gut microbiota is not essential for P-PCs and P-OPCs to positively impact health markers.

Even though many studies found that gut microbiota was a key target of the biological ability of polyphenols, most of them, it seems, just described the results of the influences of polyphenols on gut microbiota; there was little research about why or how polyphenols regulate the structure of gut microbiota. It has been proved that the bioavailability of PCs was low when administered orally and accumulates in the intestine [26]; especially for polymeric PCs, it almost cannot be absorbed and, thus, reached the colon un-metabolized, where it could directly interact with the gut microbiota. However, PCs seem not be the optimum substrate for intestinal microorganism; in particular, PCs have been found with strong bacteriostatic activity. Therefore, we hypothesized that the anti-nutrition effects of the PCs played the central role for its anti-obesity ability. Firstly, PCs could inhibit the digestion and absorption of nutrients (starch, protein, and lipids), and then these undigested nutrients transited into the colon where they could be degradation and utilization as the substrate for intestinal microorganism, then modulated the gut microbiota. This hypothesis was also proven on acarbose, an α-Glucosidase inhibitor, which could be utilized to modulate the structure of gut microbiota to increase the butyrate content through shunting starch to the colon [27].

The absorption of proanthocyanidins was highly dependent on its degrees of polymerization and the linked type, even the OPCs could be directly absorbed in vivo animal models; however, the bioavailability was low [28]. Large amounts of PCs were catabolized by gut microflora to phenolic acids and other metabolites in the colon [29]. This suggested that these metabolites of PCs might be important compounds which could also play key roles to the improvements in glucose homeostasis. Wiese et al. found that the metabolism of PCs was also associated with its DP, the formation of 5-(3′,4′-dihydroxyphenyl)-valerolactone, a main metabolites of PCs, decreased with the increasing of DP [30]. In this study, persimmon tannin (P-PCs) was a kind of highly polymerized PC (mDP = 26), and P-OPC was a mixture of degradation products (mDP = 2.8) [16]. Compared with P-OPCs, P-PCs were hardly adsorbed. Therefore, especially for P-PCs, inhibiting the digestion of may play the central role for its anti-obesity ability. Thus, we further evaluated the inhibitory activities of P-OPCs and P-PCs on the main digestion enzymes. The results indicated that both of P-OPCs and P-PCs showed strong inhibitory potentials on these enzymes, and the P-PCs showed stronger inhibitory potential compared with P-OPCs. These results were also consisted with previous studies about the tannin from sorghum, which found that the inhibitory activity of condensed tannins on α-amylase was dependent on the DP [31]. Similar results were also found in procyanidins from cocoa extracts [32]. These findings could partly explain the stronger inhibitory ability of P-PCs on α-amylase and lipase compared with P-OPCs.

5. Conclusions

In this study, the results found P-PCs with showed strong anti-obesity ability compared with P-OPCs. Compared with P-PCs, P-OPCs seems with stronger ability on improvement of insulin resistance. Furthermore, the fecal microbiota results also indicated that both of PCs and P-OPCs changed the composition of the gut microbiota and modulated specific bacteria induced by HF diet. However, it also requires further work to investigate new strategies to decrease the limitation of applications of P-PCs in food processing.

Supplementary Materials: The following supporting information can be downloaded at: https://www.mdpi.com/article/10.3390/nu14183718/s1, Table S1: The formula of D12450B and D12492.

Author Contributions: Conceptualization, K.L.; investigation, Y.Y., P.C. and X.L.; writing—original draft preparation, S.S.; writing—review and editing, K.L. All authors have read and agreed to the published version of the manuscript.

Funding: This work was supported by the Fundamental Research Funds for the Central Universities (2662020SPPY004), Scientific Research Foundation of Wuhan (WG18Q04) and Natural Science Foundation of China (31701712).

Institutional Review Board Statement: The animal study protocol was approved by the Laboratory Animal Research Center and the Ethics Committee of Huazhong Agricultural University (42816300000738).

Informed Consent Statement: Not applicable.

Data Availability Statement: All data generated or analyzed during this study are included in this published article. The data that support the findings of this study are available from the corresponding authors upon reasonable request.

Acknowledgments: We thank the Laboratory Animal Research Center for the help on the animal experiments.

Conflicts of Interest: The authors declare no conflict of interest.

References

1. Calle, E.E.; Rodriguez, C.; Walker-Thurmond, K.; Thun, M.J. Overweight, Obesity, And Mortality from Cancer In A Prospectively Studied Cohort of U.S. *Adults New Engl. J. Med.* **2003**, *348*, 1625–1638. [CrossRef] [PubMed]
2. Fruh, S.M. Obesity: Risk factors, complications, and strategies for sustainable long-term weight management. *J. Am. Assoc. Nurse Pract.* **2017**, *29*, S3–S14. [CrossRef] [PubMed]
3. Lakhani, H.V.; Sharma, D.; Dodrill, M.W.; Nawab, A.; Sharma, N.; Cottrill, C.L.; Shapiro, J.I.; Sodhi, K. Phenotypic Alteration of Hepatocytes in Non-Alcoholic Fatty Liver Disease. *Int. J. Med. Sci.* **2018**, *15*, 1591–1599. [CrossRef]
4. Dadarlat-Pop, A.; Sitar-Taut, A.; Zdrenghea, D.; Caloian, B.; Tomoaia, R.; Pop, D.; Buzoianu, A. Profile of Obesity and Comorbidities in Elderly Patients with Heart Failure. *Clin. Interv. Aging* **2020**, *15*, 547–556. [CrossRef] [PubMed]
5. Peng, J.; Jia, Y.; Hu, T.; Du, J.; Wang, Y.; Cheng, B.; Li, K. GC-(4→8)-GCG, A Proanthocyanidin Dimer from Camellia ptilophylla, Modulates Obesity and Adipose Tissue Inflammation in High-Fat Diet Induced Obese Mice. *Mol. Nutr. Food Res.* **2019**, *63*, 1900082. [CrossRef] [PubMed]
6. Anhe, F.F.; Choi, B.S.Y.; Dyck, J.R.B.; Schertzer, J.D.; Marette, A. Host-Microbe Interplay in the Cardiometabolic Benefits of Dietary Polyphenols. *Trends Endocrinol. Metab.* **2019**, *30*, 384–395. [CrossRef]
7. Tomás-Barberán, F.A.; Espín, J.C. Effect of food structure and processing on (Poly) phenol–gut microbiota interactions and the effects on human health. *Annu. Rev. Food Sci. Technol.* **2019**, *10*, 221–238. [CrossRef]
8. Li, X.; Jiao, W.; Zhang, W.; Xu, Y.; Cao, J.; Jiang, W. Characterizing the interactions of dietary condensed tannins with bile salts. *J. Agric. Food Chem.* **2019**, *67*, 9543–9550. [CrossRef]
9. Li, X.; He, C.; Song, L.; Li, T.; Cui, S.; Zhang, L.; Jia, Y. Antimicrobial activity and mechanism of Larch bark procyanidins against *Staphylococcus aureus*. *Acta Biochim. Biophys. Sin.* **2017**, *49*, 1058–1066. [CrossRef]
10. Ou, K.; Gu, L. Absorption and metabolism of proanthocyanidins. *J. Funct. Foods* **2014**, *7*, 43–53. [CrossRef]
11. Pierini, R.; Kroon, P.A.; Guyot, S.I.; Johnson, T.; Belshaw, N.J. The procyanidin-mediated induction of apoptosis and cell-cycle arrest in esophageal adenocarcinoma cells is not dependent on p21(Cip1/WAF1). *Cancer Lett.* **2008**, *270*, 234–241. [CrossRef] [PubMed]
12. Quiñones, M.; Guerrero, L.; Fernández-Vallinas, S.; Pons, Z.; Arola, L.; Aleixandre, A.; Muguerza, B. Involvement of nitric oxide and prostacyclin in the antihypertensive effect of low-molecular-weight procyanidin rich grape seed extract in male spontaneously hypertensive rats. *J. Funct. Foods* **2014**, *6*, 419–427. [CrossRef]
13. Bitzer, Z.T.; Glisan, S.L.; Dorenkott, M.R.; Goodrich, K.M.; Ye, L.; O'Keefe, S.F.; Lambert, J.D.; Neilson, A.P. Cocoa procyanidins with different degrees of polymerization possess distinct activities in models of colonic inflammation. *J. Nutr. Biochem.* **2015**, *26*, 827–831. [CrossRef] [PubMed]
14. Alves, R.G.; Mateus, N.; Freitas, V.D. Inhibition of α-amylase activity by condensed tannins. *Food Chem.* **2011**, *125*, 665–672.
15. Li, K.; Yao, F.; Du, J.; Deng, X.; Li, C.M. Persimmon Tannin Decreased the Glycemic Response through Decreasing the Digestibility of Starch and Inhibiting α-Amylase, α-Glucosidase, and Intestinal Glucose Uptake. *J. Agric. Food Chem.* **2018**, *66*, 1629–1637. [CrossRef]
16. Wang, R.; Zhang, Y.; Jia, Y.; Zhang, M.; Li, K. Persimmon Oligomeric Proanthocyanidins Exert Antibacterial Activity through Damaging the Cell Membrane and Disrupting the Energy Metabolism of *Staphylococcus aureus*. *ACS Food Sci. Technol.* **2021**, *1*, 35–44. [CrossRef]
17. Zeller, W.E.; Reinhardt, L.A.; Robe, J.T.; Sullivan, M.L.; Panke-Buisse, K. Comparison of Protein Precipitation Ability of Structurally Diverse Procyanidin-Rich Condensed Tannins in Two Buffer Systems. *J. Agric. Food Chem.* **2020**, *68*, 2016–2023. [CrossRef]
18. Shi, X.; Shang, F.; Zhang, Y.; Wang, R.; Jia, Y.; Li, K. Persimmon oligomeric proanthocyanidins alleviate ultraviolet B-induced skin damage by regulating oxidative stress and inflammatory responses. *Free Radic Res.* **2020**, *54*, 765–776. [CrossRef] [PubMed]
19. Zhu, W.; Khalifa, I.; Wang, R.; Li, C. Persimmon highly galloylated-tannins in vitro mitigated alpha-amylase and alpha-glucosidase via statically binding with their catalytic-closed sides and altering their secondary structure elements. *J. Food Biochem.* **2020**, *44*, e13234. [CrossRef]
20. Gu, L.; Kelm, M.A.; Hammerstone, J.F.; Beecher, G.; Holden, J.; Haytowitz, D.; Prior, R.L. Screening of foods containing proanthocyanidins and their structural characterization using LC-MS/MS and thiolytic degradation. *J. Agric. Food Chem.* **2003**, *51*, 7513–7521. [CrossRef] [PubMed]
21. Suo, H.; Tian, R.; Xu, W.; Li, L.; Cui, Y.; Zhang, S. Novel Catechin-Tiopronin Conjugates Derived from Grape Seed Proanthocyanidin Degradation: Process Optimization, High-Speed Counter-Current Chromatography Preparation, as Well as Antibacterial Activity. *J. Agric. Food Chem.* **2019**, *67*, 11508–11517. [CrossRef] [PubMed]
22. Paquette, M.A.; Medina Larque, S.; Weisnagel, S.J.; Desjardins, Y.; Marois, J.; Pilon, G.; Dudonne, S.; Marette, A.; Jacques, H. Strawberry and cranberry polyphenols improve insulin sensitivity in insulin-resistant, non-diabetic adults: A parallel, double-blind, controlled and randomised clinical trial. *Br. J. Nutr.* **2017**, *117*, 519–531. [CrossRef] [PubMed]
23. Stull, A.J.; Cash, K.C.; Johnson, W.D.; Champagne, C.M.; Cefalu, W.T. Bioactives in blueberries improve insulin sensitivity in obese, insulin-resistant men and women. *J. Nutr.* **2010**, *140*, 1764–1768. [CrossRef]
24. Hokayem, M.; Blond, E.; Vidal, H.; Lambert, K.; Meugnier, E.; Feillet-Coudray, C.; Coudray, C.; Pesenti, S.; Luyton, C.; Lambert-Porcheron, S. Grape Polyphenols Prevent Fructose-Induced Oxidative Stress and Insulin Resistance in First-Degree Relatives of Type 2 Diabetic Patients. *Diabetes Care* **2013**, *36*, 1454–1461. [CrossRef]
25. Wang, X.; Chen, D.; Li, Y.; Zhao, S.; Chen, C.; Ning, D. Alleviating effects of walnut green husk extract on disorders of lipid levels and gut bacteria flora in high fat diet-induced obesity rats. *J. Funct. Foods* **2019**, *52*, 576–586. [CrossRef]

26. Mendoza-Wilson, A.M.; Carmelo-Luna, F.J.; Astiazarán-García, H.; Mata-Haro, V.; Espinosa-Plascencia, A.; María, B.; Rascón-Durán, M. Absorption of dimers, trimers and tetramers of procyanidins present in apple skin by IEC-18 cell monolayers. *J. Funct. Foods* **2016**, *27*, 386–391. [CrossRef]
27. Cerqueira, F.M.; Photenhauer, A.L.; Pollet, R.M.; Brown, H.A.; Koropatkin, N.M. Starch Digestion by Gut Bacteria: Crowdsourcing for Carbs. *Trends Microbiol.* **2020**, *28*, 95–108. [CrossRef] [PubMed]
28. Wang, L.; Yamashita, Y.; Komeda, S.; Saito, A.; Ashida, H. Absorption, metabolism, distribution and faecal excretion of B-type procyanidin oligomers in mice after a single oral administration of black soybean seed coat extract. *Food Funct.* **2018**, *9*, 5362–5370. [CrossRef]
29. Masumoto, S.; Aoki, S.; Miura, T.; Shoji, T. Flavan-3-ol/Procyanidin Metabolomics in Rat Urine Using HPLC-Quadrupole TOF/MS. *Mol. Nutr. Food Res.* **2018**, *62*, e1700867. [CrossRef]
30. Wiese, S.; Esatbeyoglu, T.; Winterhalter, P.; Kruse, H.P.; Winkler, S.; Bub, A.; Kulling, S.E. Comparative biokinetics and metabolism of pure monomeric, dimeric, and polymeric flavan-3-ols: A randomized cross-over study in humans. *Mol. Nutr. Food Res.* **2015**, *59*, 610–621. [CrossRef]
31. Mkandawire, N.L.; Kaufman, R.C.; Bean, S.R.; Weller, C.L.; Jackson, D.S.; Rose, D.J. Effects of sorghum (*Sorghum bicolor* (L.) Moench) tannins on alpha-amylase activity and in vitro digestibility of starch in raw and processed flours. *J. Agric. Food Chem.* **2013**, *61*, 4448–4454. [CrossRef] [PubMed]
32. Gu, Y.; Hurst, W.J.; Stuart, D.A.; Lambert, J.D. Inhibition of Key Digestive Enzymes by Cocoa Extracts and Procyanidins. *J. Agric. Food Chem.* **2011**, *59*, 5305–5311. [CrossRef] [PubMed]

nutrients

MDPI

Article

IgE-Binding and Immunostimulating Properties of Enzymatic Crosslinked Milk Proteins as Influenced by Food Matrix and Digestibility

Sara Benedé *, Mónica Martínez-Blanco, Rosina López-Fandiño and Elena Molina *

Instituto de Investigación en Ciencias de la Alimentación (CIAL, CSIC-UAM), Nicolás Cabrera 9, 28049 Madrid, Spain

* Correspondence: s.benede@csic.es (S.B.); e.molina@csic.es (E.M.); Tel.: +34-0017900 (S.B. & E.M.)

Abstract: Dairy foods are essential in the diet, although in some susceptible individuals they may cause allergy to cow's milk proteins. Therefore, alternative methods are sought to reduce their allergenicity. Transglutaminase (TG) is widely used in dairy products mainly to improve texture. Although it has been claimed that TG can be used to modify the digestibility and allergenicity of foods, its impact within a real matrix has been rarely studied. The aim of this work was to assess the allergenic potential of crosslinked skim milk (SM), milk casein fraction (CN), and whey protein (WP). To this purpose, inhibition ELISA with sera from milk allergic patients, in vitro activation tests of mouse mast cells and splenocytes, and simulated gastrointestinal digestion assays were performed. The results showed that cross-linking increased the binding of IgE to WP, but decreased IgE-binding to SM and CN. However, no differences were observed in the ability of cross-linked proteins to induce mast cell degranulation compared to native proteins. The cross-linking of SM and CN reduced Th2 cytokine release from the splenocytes of sensitized mice. All TG-treated samples exhibited more resistance to in vitro digestion than the untreated proteins and the human IgE binding capacity after digestion was higher. In conclusion, TG treatment of milk proteins does not reduce the risk of eliciting allergic symptoms in cow's milk allergic patients.

Keywords: milk allergens; casein; whey protein; transglutaminase; allergenicity; digestion

1. Introduction

Food allergies are increasing around the world with unprecedented complexity and severity. They constitute a serious health problem that causes large public health costs and severely deteriorates the quality of life of patients. Of all known food allergies, cow's milk allergy is of special interest due to its high incidence during the first three years of life [1]. Mainstay milk allergy treatment consists of avoiding all foods containing milk proteins, nutritional counseling, as many foods cannot be consumed, and emergency treatment to reduce the allergic response. However, avoidance of milk proteins involves quite a few dietary restrictions since they are used as ingredients in many products beyond dairy foods. Furthermore, the positive contribution of milk proteins and related bioactive peptides to a healthy diet has been increasingly documented [2] and the lack of some micronutrients present in dairy foods, such as calcium, can also have other (deleterious) effects on the growth and development of children [3].

Different approaches have been used to try to produce hypoallergenic foods, taking advantage of the transformations that food components undergo during processing [4]. Among them, attention has recently been paid to protein cross-linking with transglutaminase, an enzyme that catalyzes the formation of isopeptide bonds between the amino acids glutamine and lysine in proteins, in part because it has been accepted in the European Union for the marketing of food products. Its most widespread use includes the manufacture of cheese and other dairy products, meat processing, production of edible films, and manufacture of bakery products. Studies on the crosslinking of food proteins with TG mainly evaluate its influence on physical properties of food, such as solubility, gelling, or emulsifying capacity, among others. Its contribution to the reduction of allergenicity has also been assessed [5]. However, beyond some promising results, the optimal conditions for its use have not yet been defined, nor is there consensus on its application by the food industry.

Partial hydrolysis of whey proteins or casein followed by TG-catalyzed crosslinking of the hydrolysates was proposed as a strategy to produce protein polymers with lower antibody binding properties but with improved textural and sensory properties [6,7]. Regarding the intact proteins, TG-induced polymerization was shown to decrease the antigenicity of whey protein and casein preparations [7] and TG-polymerized β-Lg gave negative prick tests in allegic subjects [8]. However, the assessment of the sensitizing and eliciting capacity of polymerized milk proteins in vivo revealed that TG-treated caseinate exhibited a somehow lower sensitizing potential but similar eliciting potential relative to its non-modified counterpart, while cross-linked β-Lg elicited a stronger allergic sensitization and a lower allergic effector response [9,10]. This different behavior was attributed to the enhanced resistance to proteolysis of cross-linked β-Lg, which redirected protein transport to PP and changed uptake and processing by antigen presenting cells [10]. Therefore, the relevance of protein cross-linking in the case of food allergy needs to be studied by taking into account the real food matrix that contains components with different resistance to processing and gastroduodenal digestion, factors which ultimately allow them to interact with the intestinal mucosa where absorption occurs [4].

The aim of this work was to assess the effect of TG-mediated crosslinking on the IgE-binding, ability to induce mast cell degranulation, and immunostimulating properties of milk in comparison with its two main protein fractions. Moreover, a physiologically relevant in vitro digestion model was used to investigate the susceptibility to digestion of the cross-linked milk proteins and the binding of human IgE to the resulting digests.

2. Materials and Methods

2.1. Crosslinking of Milk Proteins

Milk, obtained from the local farm La Colmenareña (Madrid, Spain), was skimmed by centrifugation, 3000× *g* at 4 °C for 30 min, filtered through glass wool, lyophilized, and stored at −20 °C. Commercial micellar casein Prodiet 85B (CN) was from Ingredia S.A, (Arras, France). Whey protein isolate (WP) (LACPRODAN DI-9224K) was from Arla Foods Ingredients (Sønderhøj, Denmark). Protein concentration of skim milk (SM), CN, and WP was determined by the Kjeldahl method [11]. The absence of lipopolysaccharide (LPS) was confirmed using the transfected cell line THP1-XBlue™ and the QUANTI-Blue™ assay (Invitrogen, Carlsbad, CA, USA), following the manufacturer's instructions. Protein crosslinking was performed with TG (MTG-Activa ®-MP, Ajinomoto Co., Sao Paulo, Brazil; enzymatic activity:100 U/g), a commercial powder including 1% TG and 99% maltodextrin. The enzyme was dissolved in PBS and mixed (1:1, *v/v*) with the SM, CN, and WP solutions to achieve a final protein concentration of 2.5 mg/mL and a final enzyme activity of 1, 10, 100 and 1000 U in the reaction mixture. TG cross-linking experiments were carried out at 40 °C for 1, 3, 8, and 24 h. Then, samples were heated at 90 °C for 10 min for enzyme inactivation.

2.2. Electrophoresis SDS-PAGE

Electrophoresis was performed in a Criterion™ cuvette on preform bis–tris gels (12%) at 120 V, using XT-MES as running buffer. Samples were diluted in sample buffer with 5% (*v/v*) β-mercaptoethanol to a final protein concentration of 250 µg/mL, then heated at 95 °C for 10 min. Hence, 30 µL of sample were placed into each lane. Staining was completed with Bio-Safe Coomassie G-250. All equipment and reagents were from Bio-Rad (Richmond, CA, USA).

2.3. Human IgE Binding by Inhibition ELISA

Human IgE-binding to native and TG cross-linked SM, CN, and WP and their digests was evaluated by inhibition ELISA [12], using a pool of 7 different sera from milk allergic patients with levels of IgE specific to α-LA, β-LG, and CN described in Table 1. To perform ELISA, the final protein concentration was 2.5 mg/mL and the final enzyme activity in the mixture was 100U, maintaining the reaction for 24 h.

Table 1. Sera used in the study. Specific IgE levels (kU/L) towards casein (CN), α-lactalbumin (α-LA), β-lactoglobulin (β-LG) and cow's milk of the human sera used in the inhibition ELISA and mast cell activation assay.

Serum	IgE (kU/L)			
	CN	α-LA	β-LG	Cow's Milk
1	40.0	12.1	8.3	48.2
2	16.1	5.3	7.8	23.2
3	<100	<100	<100	<100
4	73.2	29.8	5.0	54.1
5	20.5	19.2	7.6	20.1
6	>100	>100	>100	>100
7	62.7	>100	>100	>100

2.4. Bone Marrow Derived Mast Cells

Bone marrow-derived mast cells (BM-MCs) were in vitro differentiated into a mucosal-like phenotype as described previously [13]. Briefly, bone marrow cells were collected from femurs and cultured in DMEM with glucose and L-glutamine, supplemented with 10% fetal bovine serum, penicillin/streptomycin, and sodium pyruvate (Biowest, Nuaillé, France), plus stem cell factor, IL-3, IL-9 and TGF-β (Peprotech, Rocky Hill, NJ, USA), for a minimum of 4 weeks and a maximum of 8 weeks before being used. For activation through cross-linking of the IgE receptor, MCs (1×10^6 cells) were sensitized with serum from BALB/c mice orally sensitized with 5 mg of SM, CN, and WP plus 10 μg of cholera toxin (List Biologicals, Campbell, CA, USA) [14]. After washing, MCs were activated with samples in HEPES degranulation buffer. To study the activation through an IgE independent pathway, non-sensitized MCs were also stimulated with samples. β-hexosaminidase was detected by an enzymatic colorimetric assay [15]. The percentage of β-hexosaminidase release was calculated considering the total β-hexosaminidase content.

2.5. Cell Viability of Mice Splenocytes

The cell viability assay was carried under the culture conditions described below and determined by 3-(4,5-dimethylthiazol-2-yl)-2,5-diphenyltetrazolium bromide (MTT) method. Hence, 20 μL of MTT solution (5 mg/mL PBS) were added to each well. Samples were incubated for a further 4 h in dark. Then, media were removed, and 100 μL of dimethyl sulfoxide:methanol (1:1) was added to dissolve the purple insoluble MTT formazan, and the absorbance was measured at 570 nm in a Fluorostar Optima microplate-reader (BMG LabTech, Ortenberg, Alemania) using a 650 nm filter as a reference.

2.6. Cytokine Secretion of Mice Splenocytes

Splenocytes were obtained from female BALB/c mice orally sensitized with SM, CN, and WP ($n = 5$). After sacrifice, whole spleens were excised, and cell suspensions were prepared. Splenocytes were cultured in RPMI 1640 plus 10% of fetal bovine serum, L-glutamine (2 mM), penicillin (50 U/mL) and streptomycin (50 μg/mL) (all from Biowest) at 4×10^6 cells/mL in 48-well plates. Stimuli consisted of 200 μg/mL of untreated or TG cross-linked SM, CN, WP, and PBS as negative control. Control and stimulated splenocyte cultures were maintained for 72 h at 37 °C in a 5% CO_2 gassing incubator and were performed in duplicate. The supernatants were collected and kept frozen (−80 °C) until assayed for IL-4 and IL-13 quantification using mouse ELISA kits (eBioscience, San Diego, CA, USA).

2.7. In Vitro Gastrointestinal Digestion

Simulated gastrointestinal digestion of native and TG cross-linked SM, CN, and WP was performed as described previously [16]. Briefly, a gastric phase was conducted with porcine pepsin (EC 3.4.23.1, 4220 U/mg, Sigma Aldrich, St. Louis, MO, USA), at a final concentration of 182 U/mg of protein for 1 h at 37 °C adjusted to pH 2.0 (2 M HCl). The reaction was stopped by raising the pH to 7.0 (2 M NaHCO3). The duodenal phase was carried out using gastric digests by adding Corolase PP (AB Enzymes, Darmstadt, Germany) at an enzyme:substrate ratio of 1:25 (*w/w*), and incubating for 30 min at 37 °C. Stopping of the reaction was achieved by heating at 95 °C for 10 min.

2.8. RP-HPLC

RP-HPLC analysis was conducted on a Waters 600 HPLC instrument (Waters, Mildford, MA, USA) with a RP318 column (250 × 4.6 mm, 5 µm of particle size, 300Å pore size, Bio-Rad). 100 µL (2 mg/mL) of sample were injected and eluted for 60 min using 0.37% (*v/v*) trifluoroacetic acid from Scharlau Chemie (Barcelona, Spain) in double-distilled water as phase A, and 0.27% (*v/v*) trifluoroacetic acid in HPLC-grade acetonitrile (Lab-Scan, Gliwice, Poland) as phase B. The gradient was linear, increasing B in A from 0% to 60% at a flow rate of 1 mL/min during 90 min, followed by a 35-min wash with 60% and 10 min with 100% of solvent B. Detection was at 214 nm and data were processed by using Empower 2 Software from Waters.

2.9. Statistical Analysis

Statistical analyses were performed using GraphPad Prism software, version 7.0e (GraphPad). The two-tailed Student's t test was used for determining statistical significance ($p < 0.05$). Data are expressed as the mean ± SEM of 3 independent experiments.

3. Results and Discussion

3.1. Protein Polymerization

CN, WP, and SM were susceptible to TG mediated cross-linking (Figure 1). As expected, increasing TG doses and incubation times led to higher polymerization degrees. The casein fraction was the most susceptible to polymer formation due to its little secondary and tertiary structure, flexible and random-coil arrangement, and absence of any disulphide bonds in α_{S1}- and β-CNs, which allow a greater exposure of reactive groups and an increased accessibility to the enzyme [17]. The presence of protein aggregates corresponding to dimers and trimers was observed after 1 h of CN polymerization with 1 and 10 U of TG/g. After longer incubation periods, higher polymers were detected, and no presence of monomers was observed at 24 h (Figure 1a). The polymerization of β- and κ-CNs occurs earlier than that of the other milk proteins, as also observed by other authors [18]. Monomers of β-LG and α-LA were still detected after 24 h of treatment with TG, even at the highest concentration used (1000 U/g of protein) (Figure 1b). WP, due to their compact globular structures, cross-linked less efficiently [17]. Although there is no specific legislation about minimum or maximum quantities of TG allowed to be added to food products in the European Union or in the United States, most studies that addressed the use of TG in dairy foods use a maximum of 100 U/g of protein [19,20]. We evaluated the use of 1000 U/g of protein to assess whether higher enzyme activity could result in higher crosslinking efficiency, but although polymerization increased slightly compared to 100 U/g, it was not enough to convert all monomeric proteins into polymers. The dairy matrix did not change the susceptibility to cross-linking of the main milk fractions, as similar effects were observed in milk and in the isolated proteins (Figure 1c).

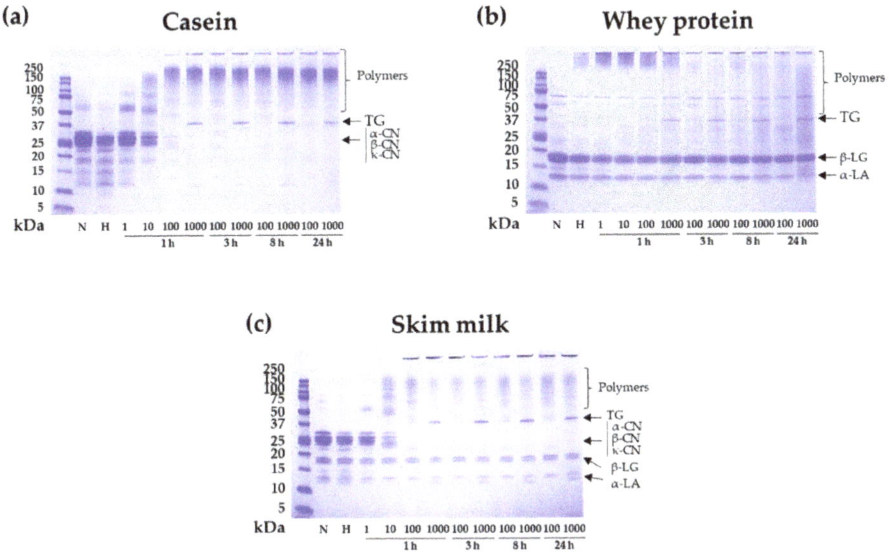

Figure 1. Electrophoretic patterns of TG treated samples. SDS-PAGE using Bis-Tris 12% acrylamide gels of (**a**) casein, (**b**) whey proteins and (**c**) skim milk after cross-linking using transglutaminase (TG) at 1, 10, 100, and 100 U of TG/g of protein during 1, 3, 8, and 24 h of incubation at 40 °C. Heating to 90 °C for 10 min to deactivate the TG was applied to the samples. Native (N) and heated (H, 90 °C for 10 min) proteins were included as control samples. α-casein = α-CN, β-casein = β-CN, k-casein = k-CN, β-lactoglobulin = β-LG and α-lactalbumin = α-LA.

3.2. Influence of Crosslinking on IgE-Binding Capacity

IgE reactivity of pooled sera from milk allergic patients from our serum library (Table 1) towards intact and polymerized samples was assessed by inhibition ELISA. Cross-linking of CN and SM led to a reduction in their IgE binding capacity. However, after treatment with TG, IgE binding to WP increased (Figure 2a), probably due to the exposure of hidden epitopes due to conformational changes in the protein [21]. Other allergenic proteins or extracts, such as ovalbumin [22], Ara h 1 [23], tropomyosin [24], β-LG [20], and wheat flour [25], have shown a reduced allergenicity after cross-linking with TG. However, peanut [26] and wheat [27] proteins showed higher IgE binding capacity after cross-linking with TG. These results highlight the importance of the substrate type in the IgE binding capacity of the TG reaction products.

The contribution of individual milk proteins to the binding of IgE capacity to SM, CN, and WP was assessed by inhibition ELISA pre-coating the plates with either β-CN, β-Lg, or α-la. In SM, the binding of IgE to β-CN was greatly reduced after treatment with TG to 3% of the IgE binding to β-CN in the untreated sample (Figure 2b). However, the specific IgE binding to β-LG and α-LA increased to 194% and 181%, respectively. This is consistent with the data obtained when specific inhibition ELISA was performed on CN and WP samples (Figure 2c,d). Different results were found by Fotschki et al., who showed that polymerization with TG decreased β-LG immunoreactivity and increased IgE binding of α-CN [20]. However, they used polyclonal rabbit antibodies, obtained from rabbits immunized against commercial purified allergens instead of human sera from naturally sensitized allergic patients. Villas-Boas et al. [19] also showed that TG crosslinking decreased the potential antigenicity of β-LG by modifying or hiding epitopes, although cross-linking was performed in the presence of cysteine.

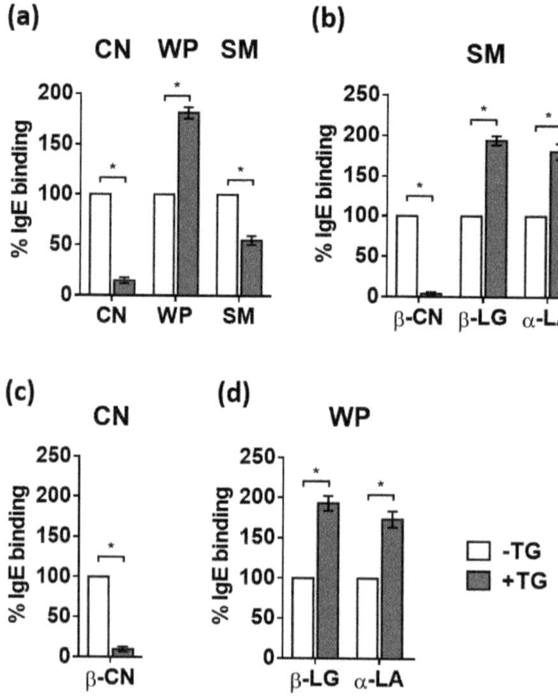

Figure 2. IgE binding to casein (CN), whey proteins (WP) and skim milk (SM), either untreated (−TG, white bars) or treated with transglutaminase (+ TG, grey bars), estimated by inhibition ELISA with sera from milk allergic patients (Table 1). (**a**) IgE binding to CN, WP, and SM, (**b**) specific IgE binding to β-casein (β-CN), β-lactoglobulin (β-LG), and α-lactalbumin (α-LA) in SM; (**c**) specific IgE binding to β-CN in CN, and (**d**) specific IgE binding to β-LG and α-LA in WP. The proteins used for pre-coating the plates are indicated on the x-axes. Data are expressed as means ± SEM of 3 different experiments. Statistically significant differences ($p < 0.05$) are indicated by asterisks.

Because mast cells are the main effector cells of the allergic response, we assessed the ability of TG-treated SM, CN, and WP to activate them through an IgE dependent pathway using BM-MCs (Figure 3). Cells were primed with sera from mice sensitized to SM, CN or WP and then treated with specific antigens or PBS. No differences were observed in the percentage of β-hexosaminidase released by BM-MC after incubation with TG-treated SM, CN, and WP compared to cells stimulated with the untreated proteins. To rule out that cell activation occurred through a non-IgE-mediated pathway, we confirmed that neither the protein samples treated with TG nor TG alone induced degranulation in non-sensitized cells. These data would indicate that, although the IgE-binding capacity of milk proteins is affected by the action of TG, the effector cellular response may not be altered. A similar behavior was observed by Nakamura et al. [28], who observed a pronounced increase in the IgE reactivity of a hydrolyzed wheat protein treated with TG without affecting the IgE response of patients with wheat allergy.

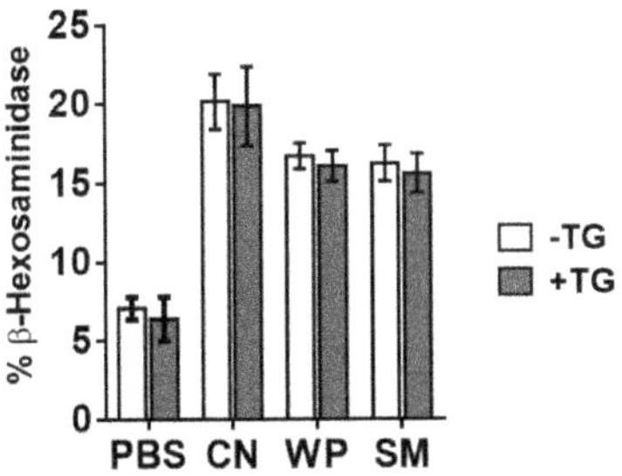

Figure 3. Mast cell degranulation capacity. Measurement of β-hexosaminidase release induced by casein (CN), whey proteins (WP) and skim milk (SM), either untreated (white bars) or treated with TG (grey bars), after sensitization of bone marrow derived mast cells with sera from CN, WP and SM allergic mice, respectively. Data are expressed as mean ± SEM of 3 different experiments.

3.3. Immunostimulating Properties of Crosslinked Milk Proteins

The immunostimulating properties of TG-treated SM, CN, and WP were evaluated in splenocytes from SM, CN, and WP sensitized BALB/c mice, respectively (Figure 4). Stimulation of spleen cells with untreated CN, WP, and SM led to an increased Th2 cytokine production compared with control cells (PBS). Cytokine secretion from cells stimulated with TG and PBS was similar (not shown), indicating an antigen specific stimulation. Cross-linking of SM and CN reduced the release of IL-4 from the splenocytes to a level similar to that of control cells. IL-13 secretion was also reduced, although it was not significantly different with respect to the control cells. In contrast, IL-4 and IL-13 production in splenocytes stimulated with WP was similar regardless of whether WP was cross-linked or not. Th2 cytokines (IL-4, IL-5, and IL-13) secreted by differentiated and expanded T-helper lymphocytes favor the activation of B lymphocytes that differentiate into plasma cells, producing large amounts of IgE antibodies, whose activation on the surface of basophils and mast cells triggers the allergic response [29]. The lower levels of IL-4 and IL-13 detected in the supernatants of splenocytes incubated with TG-treated SM and CN, but not WP, may indicate that crosslinked SM and CN exert positive immunosuppressive effects.

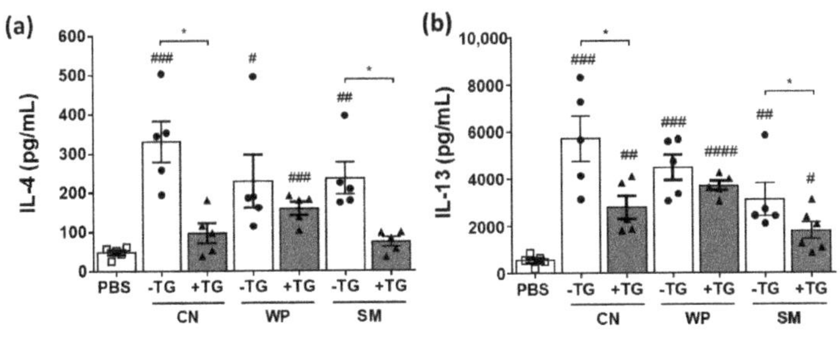

Figure 4. Stimulation of splenocytes. Th2 cytokines (**a**) IL-4 and (**b**) IL-13 released from cultures of spleen cells from casein (CN), whey protein (WP), and skim milk (SM) sensitized BALB/c mice (n = 5) stimulated with CN, WP, and SM, either untreated (white bars) or treated with TG (grey bars). Pounds and asterisks indicate, respectively, statistically significant differences with respect to cells stimulated with PBS or between both experimental groups. Data are expressed as mean ± SEM. # and * $p < 0.05$, ## $p < 0.01$, ### $p < 0.001$, #### $p < 0.0001$.

Some studies have determined the influence of crosslinking on the allergenic properties of proteins through the analysis of the IgE binding capacity by ELISA, Western blotting, or degranulation tests using basophil and mast cell cultures [30–32], but very few have determined their impact on other immune cells. Bai et al. [33] observed that TG cross-linked tofu proteins reduced the levels of Th2 cytokines released by spleen cells from mice allergic to tofu protein cultured in vitro compared to native proteins. Similarly, ovalbumin and egg white subjected to high-pressure treatment and crosslinking (simultaneously or thereafter) induced a lower IL-5 and IL-13 release from splenocytes [22].

None of the samples analyzed in this study exhibited spleen cell cytotoxicity after a 72 h-culture (Figure 5), indicating that the inhibitory effects on cytokine release caused by TG were due to a lower reactivity of the cross-linked proteins. Other in vivo and in vitro studies of toxicity of microbial TG have revealed safety concerns about its food applications [34] and therefore more research is need in this regard.

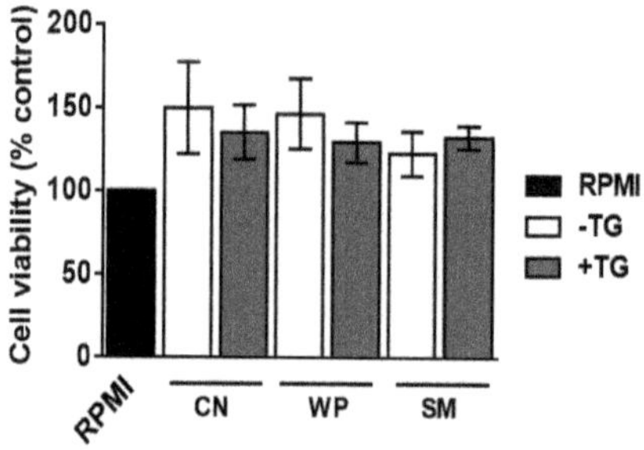

Figure 5. Cytotoxicity assay. The MTT method was used to determine the viability of spleen cells from casein (CN), whey protein (WP), and skim milk (SM) sensitized BALB/c mice cultured with either untreated (white bars) or TG cross-linked CN, WP, and SM (grey bars).

3.4. Impact of Crosslinking on Digestibility of Milk Proteins

The resistance to in vitro gastric digestion of TG treated SM, CM, and WP was higher than that of the native proteins, and this was especially noticeable during the first minutes of pepsin treatment (Figure 6). Resistance to digestion could be used as a predictive factor in assessing the allergenic potential of a protein [35] and, in this respect, it is essential to evaluate the digestion process of proteins within the food matrix in which they are usually found [20]. As shown in Figure 6, digestion with pepsin did not lead to complete hydrolysis, as there was intact protein in all cases, while SM and CN were almost completely hydrolyzed with Corolase PP. In the case of WP, intact protein was still observed after 30 min of simulated duodenal digestion in both native and TG-treated proteins, although the peak corresponding to intact proteins was higher in the cross-linked sample.

The peptide profile generated during digestion in the gastric and duodenal phases was similar in the native samples and in those treated with TG. Similar results were found by Fotschki et al., who observed that TG influenced the gastric digestibility of cow and horse milk, slowing digestion of the proteins [20]. Cross-linked β-LG was found to be more resistant to simulated gastrointestinal digestion [10]. In contrast, polymerization of chemically denatured β-LG with TG was reported to facilitate the action of gastrointestinal enzymes [32]. Similarly, Havenaar et al. [36] indicated that caseinate treated with TG was more digestible than the untreated protein. These discrepancies may be due to the use of different polymerization conditions and in vitro digestion protocols, as it is well known that the rate of digestion of polymerized proteins depends on the type of structure generated [37].

As expected, in vitro gastrointestinal digestion significantly reduced the IgE-binding capacity of SM, CN, and WP (Table 2). The concentration (μg/mL) for 50% of the maximal IgE binding, also known as EC50, of SM and CN treated with TG after gastric and duodenal digestion was approximately an order of magnitude higher than that of the digests for the native samples and similar results were obtained for WP samples. These data are in agreement with the lower digestibility of samples cross-linked with TG observed by RP-HPLC (Figure 6).

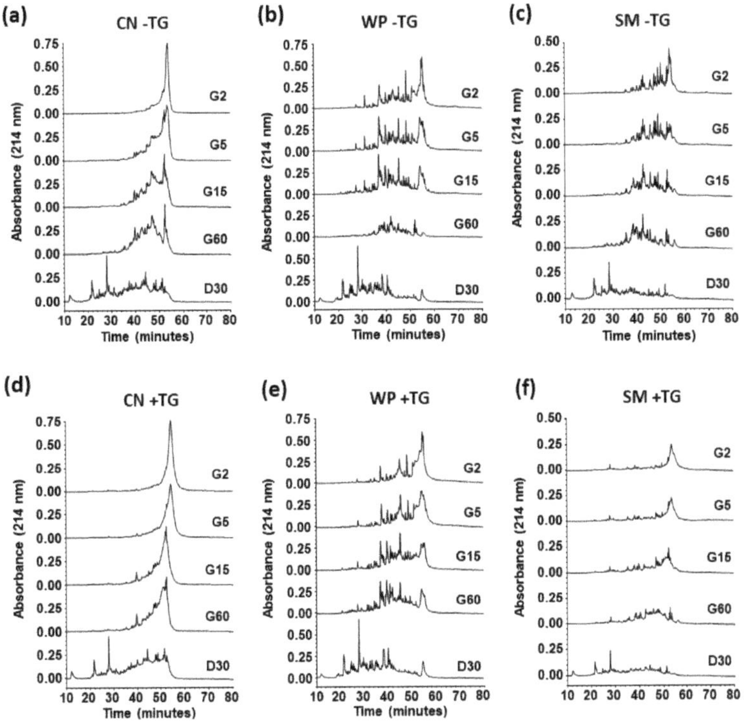

Figure 6. Digestibility of TG cross-linked milk proteins. RP-HPLC patterns of untreated (−TG; **a–c**) and TG cross-linked (+TG; **d–f**) casein (CN; **a**,**d**), whey proteins (WP; **b**,**e**), and skim milk (SM; **c**,**f**), after 2 (G2), 5 (G5), 15 (G15), and 60 (G60) min of simulated gastric digestion with pepsin, and subsequent duodenal digestion with Corolase PP for 30 min (D30).

Table 2. IgE binding to untreated and TG cross-linked slim milk (SM), casein (CN), and whey proteins (WP), before and after gastric and duodenal digestion, as estimated by inhibition ELISA with sera from milk allergic patients (EC50, μg/mL).

	Undigested		**Gastric**		**Duodenal**	
	−TG	**+TG**	**−TG**	**+TG**	**−TG**	**+TG**
Casein	0.11	0.79	35.50	4.96	289.23	21.19
	(+/−0.01)	(+/−0.15)	(+/−5.81)	(+/−0.42)	(+/−14.48)	(+/−3.92)
Whey protein	0.74	0.41	1042.59	87.33	1236.17	49.45
	(+/−0.03)	(+/−0.01)	(+/−107.98)	(+/−10.17)	(+/−112.38)	(+/−7.05)
Skim milk	0.39	0.72	99.26	23.81	31,470.27	9188.28
	(+/−0.01)	(+/−0.06)	(+/−12.32)	(+/−3.48)	(+/−3567.51)	(+/−442.36)

Values are the mean of 3 independent experiments (standard deviation in brackets).

4. Conclusions

TG was effective in crosslinking SM, CN, and WP, forming high molecular weight polymers, although a substantial amount of unmodified monomeric serum proteins remained. Although the binding of human IgE to β-CN was significantly reduced and that to β-LG and α-LA was increased after treatment of SM, CN, and WP with TG, no differences were observed in either case in the ability to induce mast cell degranulation compared to the untreated proteins. SM and CN subjected to cross-linking induced a lower release of

Th2 cytokines from splenocytes of mice sensitized, respectively, to SM and CN, but WP treated with TG did not modify the production of IL-4 and IL-13 with respect to the native protein in spleen cells from WP-sensitized mice. TG-treated SM, CM, and WP were more resistant to digestion than the untreated proteins and, consequently, the binding of human IgE to their digests was also higher. These results question the use of crosslinking with transglutaminase to obtain hypoallergenic products.

Author Contributions: Conceptualization: S.B. and E.M.; methodology: M.M.-B. and S.B.; validation: S.B. and R.L.-F.; formal analysis: S.B.; investigation: M.M.-B., S.B., E.M. and R.L.-F.; resources: S.B., E.M. and R.L.-F.; data curation: S.B., E.M. and R.L.-F.; writing—original draft preparation: S.B. and E.M.; writing—review and editing, all authors.; visualization: E.M. and S.B.; supervision, all authors; project administration: E.M. and R.L.-F.; funding acquisition: E.M. and R.L.-F.; final approval of the version, all authors. All authors have read and agreed to the published version of the manuscript.

Funding: This work was supported by the Ministerio de Ciencia, Innovación y Universidades (project AGL2017-88964-R) and Ministerio de Ciencia e Innovación through a project (PID2021-123236OB-I00) and a Juan de la Cierva Incorporación contract (SB) (IJCI-2017-31345 B).

Institutional Review Board Statement: The study was conducted according to the guidelines of the Declaration of Helsinki, following the European legislation (Directive 2010/63/EU) and approved by the Institutional Review Board of Dirección General de Agricultura, Ganadería y Alimentación from Comunidad de Madrid (PROEX 286.8/20).

Informed Consent Statement: Not applicable.

Data Availability Statement: Not applicable.

Acknowledgments: CSIC Interdisciplinary Thematic Platform (PTI) Alergias alimentarias. (PTI-FOODALLERGY).

Conflicts of Interest: The authors declare no conflict of interest.

References

1. Flom, J.D.; Sicherer, S.H. Epidemiology of cow's milk allergy. *Nutrients* **2019**, *11*, 1051. [CrossRef]
2. McGregor, R.; Poppitt, S.D. Milk proteins and human health. In *Milk Proteins: From Expression to Food*; Singh, H., Boland, M., Thompson, A., Eds.; Academic Press: London, UK; Waltham, MA, USA; San Diego, CA, USA, 2014; pp. 651–669.
3. Vassilopoulou, E.; Christoforou, C.; Andreou, E.; Heraclides, A. Effects of food allergy on the dietary habits and intake of primary schools' Cypriot children. *Eur. Ann Allergy Clin. Immnunol.* **2017**, *49*, 181–185. [CrossRef]
4. Verhoeckx, K.C.M.; Vissers, Y.M.; Baumert, J.L.; Faludi, R.; Feys, M.; Flanagan, S.; Herouet-Guicheney, C.; Holzhauser, T.; Shimojo, R.; van der Bolt, N.; et al. Food processing and allergenicity. *Food Chem. Toxicol.* **2015**, *80*, 223–240. [CrossRef]
5. Gharibzahedi, S.M.T.; Koubaa, M.; Barba, F.J.; Greiner, R.; George, S.; Roohinejad, S. Recent advances in the application of microbial transglutaminase crosslinking in cheese and ice cream products: A review. *Int. J. Biol. Macromol.* **2017**, *107*, 2364–2374. [CrossRef]
6. Sabadin, I.S.; Villas-Boas, M.B.; de Lima Zollner, R.; Netto, F.M. Effect of combined treatment of hydrolysis and polymerization with transglutaminase on β-lactoglobulin antigenicity. *Eur. Food Res. Technol.* **2012**, *235*, 801–809. [CrossRef]
7. Damodaran, S.; Li, Y. A two-step enzymatic modification method to reduce immuno-reactivity of milk proteins. *Food Chem.* **2017**, *237*, 724–732. [CrossRef]
8. Olivier, C.E.; Lima, R.P.; Pinto, D.G.; Santos, R.A.; Silva, G.K.; Lorena, S.L.; Villas-Boas, M.B.; Netto, F.M.; de Lima Zollner, R. In search of a tolerance-induction strategy for cow's milk allergies: Significant reduction of beta-lactoglobulin allergenicity via transglutaminase/cysteine polymerization. *Clinics* **2012**, *67*, 1171–1179. [CrossRef]
9. Van Esch, B.C.; Gros-van Hest, M.; Westerbeek, H.; Garssen, J. Sensitizing capacity and allergenicity of enzymatically cross-linked sodium caseinate in comparison to sodium caseinate in a mouse model for CMA. *Toxicol. Lett.* **2013**, *218*, 50–55. [CrossRef] [PubMed]
10. Stojadinovic, M.; Pieters, R.; Smit, J.; Cirkovic Velickovic, T. Cross-linking of β-lactoglobulin enhances allergic sensitization through changes in cellular uptake and processing. *Toxicol. Sci.* **2014**, *140*, 224–235. [CrossRef] [PubMed]
11. Sosulski, F.W.; Imafidon, G.I. Amino-acid-composition and nitrogen-to-protein conversion factors for animal and plant foods. *J. Agric. Food Chem.* **1990**, *38*, 1351–1356. [CrossRef]
12. Jiménez-Sáiz, R.; Martos, G.; Carrillo, W.; López-Fandiño, R.; Molina, E. Susceptibility of lysozyme to in-vitro digestion and immunoreactivity of its digests. *Food Chem.* **2011**, *127*, 1719–1726. [CrossRef]
13. Benedé, S.; Cody, E.; Agashe, C.; Berin, M.C. Immune characterization of bone marrow-derived models of mucosal and connective tissue mast cells. *Allergy Asthma Clin. Immunol. Res.* **2018**, *10*, 268–277. [CrossRef] [PubMed]

14. Pablos-Tanarro, A.; Lopez-Expósito, I.; Lozano-Ojalvo, D.; López-Fandiño, R.; Molina, E. Antibody production, anaphylactic signs, and T-cell responses induced by oral sensitization with ovalbumin in BALB/c and C3H/HeOuJ mice. *Allergy Asthma Immnunol. Res.* **2015**, *8*, 239–245. [CrossRef] [PubMed]
15. Benedé, S.; Berin, M.C. Mast cell heterogeneity underlies different manifestations of food allergy in mice. *PLoS ONE* **2018**, *13*, e0190453. [CrossRef] [PubMed]
16. Benedé, S.; Lopez-Exposito, I.; Gimenez, G.; Grishina, G.; Bardina, L.; Sampson, H.A.; Lopez-Fandiño, R.; Molina, E. Mapping of IgE epitopes in in vitro gastroduodenal digests of beta-lactoglobulin produced with human and simulated fluids. *Food Res. Int.* **2014**, *62*, 1127–1133. [CrossRef]
17. Özrenk, E. The use of transglutaminase in dairy products. *Int. J. Dairy Technol.* **2006**, *59*, 1–7. [CrossRef]
18. Hsieh, J.F.; Pan, P.H. Proteomic profiling of microbial transglutaminase-induced polymerization of milk proteins. *J. Dairy Sci.* **2012**, *95*, 580–589. [CrossRef]
19. Villas-Boas, M.B.; Vieira, K.P.; Trevizan, G.; Zollner, R.D.; Netto, F.M. The effect of transglutaminase-induced polymerization in the presence of cysteine on b-lactoglobulin antigenicity. *Int. Dairy J.* **2010**, *20*, 386–392. [CrossRef]
20. Fotschki, J.; Wróblewska, B.; Fotschki, B.; Kalicki, B.; Rigby, N.; Mackie, A. Microbial transglutaminase alters the immunogenic potential and cross-reactivity of horse and cow milk proteins. *J. Dairy Sci.* **2019**, *103*, 2153–2166. [CrossRef]
21. Duarte, N.M.; Cruz, A.L.; Silva, D.C.; Cruz, G.M. Intake of whey isolate supplement and muscle mass gains in young healthy adults when combined with resistance training: A blinded randomized clinical trial (pilot study). *J. Sport. Med. Phys. Fit.* **2020**, *60*, 75–84. [CrossRef]
22. Ma, X.; Lozano-Ojalvo, D.; Chen, H.; Lopez-Fandiño, R.; Molina, E. Effect of high pressure-assisted crosslinking of ovalbumin and egg white by transglutaminase on their potential allergenicity. *Innov. Food Sci. Emerg. Technol.* **2015**, *29*, 143–150. [CrossRef]
23. Tian, Y.; Liu, C.; Xue, W.; Wang, Z. Crosslinked recombinant-Ara h 1 catalyzed by microbial transglutaminase: Preparation, structural characterization and allergic assessment. *Foods* **2020**, *9*, 1508. [CrossRef] [PubMed]
24. Fu, L.; Ni, S.; Wang, C.; Wang, Y. Transglutaminase-catalysed cross-linking eliminates Penaeus chinensis tropomyosin allergenicity by altering protein structure. *Food Agric. Immnunol.* **2019**, *30*, 296–308. [CrossRef]
25. Leszczyńska, J.; Łącka, A.; Bryszewska, M. The use of transglutaminase in the reduction of immunoreactivity of wheat flour. *Food Agric. Immnunol.* **2006**, *17*, 105–113. [CrossRef]
26. Meng, S.; Tan, Y.; Chang, S.; Li, J.; Maleki, S.; Puppala, N. Peanut allergen reduction and functional property improvement by means of enzymatic hydrolysis and transglutaminase crosslinking. *Food Chem.* **2020**, *302*, 125186. [CrossRef]
27. Palosuo, K.; Varjonen, E.; Nurkkala, J.; Kalkkinen, N.; Harvima, R.; Reunala, T.; Alenius, H. Transglutaminase-mediated cross-linking of a peptic fraction of ω-5 gliadin enhances IgE reactivity in wheatdependent, exercise-induced anaphylaxis. *J. Allergy Clin. Immnunol.* **2003**, *111*, 1386–1392. [CrossRef]
28. Nakamura, R.; Nakamura, R.; Sakai, S.; Urisu, A.; Fukutomi, Y.; Teshima, R. Tissue transglutaminase generates deamidated epitopes on gluten, increasing reactivity with hydrolyzed wheat protein-sensitized IgE. *J. Allergy Clin. Immnunol.* **2013**, *132*, 1436–1438. [CrossRef]
29. Benedé, S.; Blázquez, A.B.; Chiang, D.; Tordesillas, L.; Berin, M.C. The rise of food allergy: Environmental factors and emerging treatments. *EBioMedicine* **2016**, *7*, 27–34. [CrossRef]
30. Stanic, D.; Monogioudi, E.; Ercili, D.; Radosavljevic, J.; Atanaskovic-Markovic, M.; Vuckovic, O.; Raija, L.; Mattinen, M.; Buchert, J.; Cirkovic Velickovic, T. Digestibility and allergenicity assessment of enzymatically crosslinked β-casein. *Mol. Nutr. Food Res.* **2010**, *54*, 1273–1284. [CrossRef]
31. Olivier, C.E.; Villas-Boas, M.B.; Netto, F.M.; Zollner, R.D.L. Allergenicity of Bos d 5 in children with cow's milk allergyis reduced by transglutaminase polymerization. *Pediatr. Allergy Immunol. Pulmonol.* **2012**, *25*, 30–33. [CrossRef]
32. Villas-Boas, M.B.; Benedé, S.; Zollner, R.D.; Netto, F.M.; Molina, E. Epitope resistance to the simulated gastrointestinal digestion of beta-lactoglobulin submitted to two-step enzymatic modification. *Food Res. Int.* **2015**, *72*, 191–197. [CrossRef]
33. Bai, J.; Hui, J.; Lu, Q.; Yang, A.; Yuan, J.; Gao, J.; Wu, Z.; Li, X.; Tong, P.; Chen, H. Effect of transglutaminase cross-linking on the allergenicity of tofu based on a BALB/c mouse model. *Food Func.* **2020**, *11*, 404–413. [CrossRef] [PubMed]
34. De Souza, C.F.V.; Venzke, J.G.; Rosa, R.M.; Henriques, J.A.P.; Dallegrave, E.; Flôres, S.H.; Ayub, M.A.Z. Toxicological evaluation for food applications of transglutaminase from a newly isolated Bacillus circulans BL32. *Am. J. Food Technol.* **2011**, *6*, 460–471. [CrossRef]
35. Verhoeckx, K.; Bøgh, K.L.; Dupont, D.; Egger, L.; Gadermaier, G.; Larré, C.; Mackie, A.; Menard, O.; Adel-Patient, K.; Picariello, G.; et al. The relevance of a digestibility evaluation in the allergenicity risk assessment of novel proteins. Opinion of a joint initiative of COST action ImpARAS and COST action INFOGEST. *Food Chem. Toxicol.* **2019**, *129*, 405–423. [CrossRef]
36. Havenaar, R.; de Jong, A.; Koenen, M.E.; van Bilsen, J.; Janssen, A.M.; Labij, E.; Westerbeek, H.J. Digestibility of transglutaminase cross-linked caseinate versus native caseinate in an in vitro multicompartmental model simulating young child and adult gastrointestinal conditions. *J. Agric. Food Chem.* **2013**, *61*, 7636–7644. [CrossRef]
37. Macierzanka, A.; Böttger, F.; Lansonneur, L.; Groizard, R.; Jean, A.S.; Rigby, N.M.; Cross, K.; Wellner, N.; Mackie, A.R. The effect of gel structure on the kinetics of simulated gastrointestinal digestion of bovine β-lg. *Food Chem.* **2012**, *134*, 2156–2163. [CrossRef]

nutrients

Article

Release of Bioactive Peptides from *Erythrina edulis* (*Chachafruto*) Proteins under Simulated Gastrointestinal Digestion

Jessica L Correa [1], José Edgar Zapata [1] and Blanca Hernández-Ledesma [2,*]

1 Nutrition and Food Technology Group, Universidad de Antioquia, Medellin 050010, Colombia
2 Development and Innovation in Alternative Proteins Group, Institute of Food Science Research (CIAL, CSIC-UAM, CEI UAM+CSIC), Nicolás Cabrera, 28049 Madrid, Spain
* Correspondence: b.hernandez@csic.es

Abstract: The estimated and concerning rise in world population over the next few years and the consequent increase in food demand will lead to a deterioration in global food security. To avoid or reduce this world crisis, informed and empowered consumers are turning to sustainable and nutrient-rich foods that substitute animal products, also reducing their associated environmental impact. Moreover, due to the demonstrated influence of diet on the risk of high incidence and mortality of noncommunicable diseases, the current established food pattern is focused on the consumption of foods that have functionality for health. Among these new foods, traditional and underutilized plants are gaining interest as alternative protein sources providing nutritional and biological properties. In this work, the potential of *Erythrina edulis* (*chachafruto*) proteins as a source of multifunctional peptides after transit through the gastrointestinal tract has been demonstrated, with antioxidant and immunostimulating effects in both biochemical assays and cell culture. While low molecular weight peptides released during the digestive process were found to be responsible for protection against oxidative stress mediated by their radical scavenging activity, high molecular weight peptides exerted immunostimulating effects by upregulation of immunoresponse-associated biomarkers. The findings of this study support the promising role of chachafruto proteins as a new antioxidant and immunostimulatory ingredient for functional foods and nutraceuticals.

Keywords: *Erythrina edulis*; simulated gastrointestinal digestion; multifunctional peptides; immunomodulatory activity

1. Introduction

In the last few years, the estimated rising of the world population, the consequent deterioration in global food security, and the concern about the environmental impact of food systems have made international organizations, governments, industries, and societies recommend a dietary shift to reduce the intake of calories from animal sources and increase the consumption of sustainable, nutrient-rich, and calorically efficient products [1]. Moreover, due to the worldwide rise in incidence and prevalence of noncommunicable diseases (NCDs) and the demonstrated impact of diet on the risk of these disorders, the current established food pattern is focused on the consumption of foods that have functionality for health. Because of the interest gained in plant proteins as part of a healthy and sustainable diet, research on traditional and underutilized plants as a source of highly nutritional and bioactive proteins has intensified in the last years [2,3].

Erythrina edulis (*chachafruto*) is a legume endemic to some South American countries with a wide range of nutritional and medicinal uses [4]. Its protein content (18–24%) and quality is superior to that reported for other legumes and similar to that of egg proteins [5]. However, although the protein content is high, the data available on the potential of these proteins as a source of bioactive peptides are still scarce. Preliminary studies performed in our laboratory have described the in vitro antioxidant, antidiabetic, and antihypertensive activity of chachafruto protein-derived hydrolysates by microbial or gastric enzymes [6,7]. However, to the best of our knowledge, there are no data on the potential of gastrointestinal digestion simulating physiological conditions to hydrolyze *E. edulis* proteins and release multifunctional peptides.

Simulated gastrointestinal digestion has become one of the most widely studied approaches to generate bioactive peptides from food proteins because it is fast, inexpensive, safe, and does not include ethical restrictions [8,9]. In vitro digestion tests allow releasing peptides encrypted within the protein matrix, simulating the process naturally occurring into the human body. Moreover, it allows understanding of how dietary proteins are degraded under physiological conditions [10]. A static gastrointestinal digestion was successfully conducted to degrade legume proteins such as yellow peas pinto beans and release peptides with antioxidant and/or ACE inhibitory activity [11,12].

Although a majority of the studies have focused on antihypertensive, antimicrobial, antidiabetic or antiproliferative peptides [13,14], the recent research on immunomodulatory peptides with positive impact on the host defense has gained interest [15,16]. However, peptides with immunomodulatory effects represent the most diverse and complex group, being needed for screening different targets (oxidative status, and inflammatory and immunoresponse-associated biomarkers) to confirm their activity. The variety of conditions resulting from compromised function of the immune system and the wide range of bioactivities that immunomodulatory peptides can exert make them promising ingredients for novel functional foods and/or nutraceuticals [17].

Thus, this work aimed at investigating the potential of gastrointestinal digestion simulating physiological conditions to release multifunctional peptides, focusing on their radical scavenging capacity and their protective effects on oxidative stress and immunoresponse-associated biomarkers.

2. Materials and Methods

2.1. Materials

E. edulis (chachafruto) seeds were collected in Rioblanco–Tolima (Colombia). Alcalase 2.4 U/g protein from *Bacillus licheniformis* was obtained from Novozymes (Bagsvaerd, Denmark). Bovine seroalbumin (BSA), 2,2′-azino-bis(3-ethylbenzothiazoline-6-sulfonic acid) (ABTS), 6-hydroxy-2,5,7,8-tetramethylchroman-2-carboxylic acid (trolox), disodium fluorescein (FL), 2,2′-azobis (2-amidinopropane dihydrochloride) (AAPH), fetal bovine serum (FBS), Dulbecco's Modified Eagle's Medium/High Modified (DMEM), L-glutamine solution, sodium pyruvate, non-essential amino acids (NEAA), gentamicin, 3-[4,5-dimethylthiazol-2-yl]-2,3-diphenyl tetrazolium bromide (MTT) were acquired from Sigma-Aldrich (St. Louis, MO, USA). All other reagents were of analytical grade.

2.2. Seed Flour Analysis

Seeds were cleaned, peeled, cut into tiny pieces, dried at 40 °C, ground in a domestic mill, and filtered using mesh number 60. The proximate composition of the resulting homogenous flour sample was analyzed following the AOAC methods [18] for moisture (934.01), fat (930.09), ash (930.05), crude fiber (934.10), and protein (978.04). The conversion factor to calculate the protein value was 6.25.

2.3. Protein Concentrate from Chachafruto Seeds (CPC)

Chachafruto flour was suspended in water (10%, *w/v*), its pH was adjusted to 10.0 by adding 1 N NaOH, and the suspension was magnetically stirred for 2 h at 60 °C, and centrifugated at 4500× *g* for 20 min at 25 °C. The pH of the supernatants was adjusted to 4.5 with 1 N HCl and the suspensions were left at 4 °C overnight. The precipitated proteins were collected by centrifugation at 10,000× *g* for 20 min at 4 °C, resuspended in water, lyophilized, and stored at −20 °C until further analyses. The amino acid content of the chachafruto protein concentrate (CPC) was analyzed by cation exchange chromatography using a Biochrom 30 series Amino Acid Analyser (Biochrom, Cambridge, UK) after automatic post-column derivatization of samples with ninhydrin and measurement of the absorbance at 570 nm. The previous hydrolysis of the samples was performed with 6 M HCl for 21 h at 110 °C. The results were expressed as mean of two replicates (g of amino acid/100 g protein).

The protein profiles of CPC, their digests, and fractions were analyzed by SDS-PAGE using 12% polyacrylamide gels (Bis-Tris CriterionTM XT Precast Gel, Bio-Rad, Hercules, CA, USA). After mixing the samples with sample buffer [60 mM Tris-HCl pH 6.8, 25% glycerol (*v/v*), 2% sodium dodecyl sulfate (SDS) (p/v), 14.4 mM 2-mercaptoethanol, and 0.1% 2-bromophenol (p/v)], they were heated for 5 min at 100 °C and cooled to room temperature before loading onto gels (50 μg of protein/sample). The analysis was performed in a Criterion automated system (Bio-Rad) using the XT MES Running Buffer 20X (Bio-Rad). The electrophoretic migration was carried out with a voltage of 150 V for 45 min. The gels were stained with Coomassie Blue for 60 min and washed with a 10% acetic acid −10% methanol solution for 12 h. Precision Plus Protein Standard Unstained (Bio-Rad) was used as molecular weight protein standard.

2.4. Simulated Gastrointestinal Digestion

CPC was digested according to the in vitro harmonized protocol [19] with some modifications. Briefly, 0.5 g of CPC was dissolved in 5 mL of salivary fluid (pH 7.0, 37 °C) for 5 min. Then, 4 mL of simulated gastric juice (pH 3.0, 37 °C) containing pepsin from porcine gastric mucosa (2000 U/mL of digest, EC 3.4.23.1, Sigma-Aldrich) was added and incubated at 37 °C for 120 min. After adjusting the pH of gastric digests (GDCPC) to 7.0 with 1 M NaOH, and intestinal phase started after addition of simulated intestinal fluid composed of porcine pancreas pancreatin (100 U trypsin activity/mL of final mixture, Sigma-Aldrich) and porcine bile extract (10 mM in the final mixture, Sigma-Aldrich). Digestions (in duplicate) were performed at 37 °C in an orbital shaker at 150 rpm. After 120 min incubation, samples were heated at 80 °C for 15 min to inactive enzymes and obtain the gastrointestinal digest (GIDCPC). Samples were freeze-dried and kept at −20 °C until their analysis. A digestion blank (DB) consisting of the enzyme mixture without CPC was prepared.

CPC, GDCPC, and GIDCPC were subjected to ultrafiltration through hydrophilic 10,000 and 3000 Da cutoff membranes (Merck KGaA, Darmstadt, Germany). Fractions of <3 kDa, 3–10 kDa and >10 kDa were lyophilized and kept at −20 °C until use. The bicinchoninic acid method (BCA) (Pierce, Rockford, IL, USA) was used to measure the peptide content of digests and fractions, using BSA as standard protein.

Analysis of peptide mass distribution in GDCPC and GIDCPC was performed by MALDI-TOF mass spectrometry (MS), following the protocol previously described [20]. Samples were spotted on an Anchorchip target (Bruker Daltonik GmbH, Bremen, Germany) with α-cyano-4-hydroxycinnamic acid (α-CHCA) matrix in acetonitrile/water (30:70) containing 0.1% trifluoroacetic acid (TFA) and analyzed on a Bruker Autoflex Speed® (Bruker Daltonik). Mass spectra were obtained in positive reflectron mode by collecting 1000 laser pulses on average. Calibration was made with the Peptide Calibration Standard I and II (Bruker Daltonik).

2.5. Antioxidant Activity by Biochemical Assays

The ABTS•+ scavenging activity was determined according to the enhanced protocol described by Re et al. [21]. 180 μL of diluted ABTS•+ solution and 20 μL of either PBS (blank), trolox (25–200 μM) (standard) or sample (at different concentrations) was mixed and incubated for 5 min at room temperature. After that time, the absorbance was measured at 734 nm in a Synergy HTX microplate reader (BioTek Instruments, Inc., Winooski, VT, USA). The trolox equivalent antioxidant capacity (TEAC) was calculated dividing the gradient of the plot of the percentage inhibition of absorbance versus protein concentration of sample by the gradient of the plot for trolox. The results were expressed as μmol trolox equivalent (TE)/mg of protein.

The oxygen radical absorbance capacity (ORAC) was determined according to the protocol described by Hernández-Ledesma et al. [22] with some modifications. A volume of 20 μL of either PBS (blank), trolox (0.2–1.6 nmol) or sample (at different concentrations) was mixed with 120 μL of FL (117 nM) and incubated at 37 °C for 10 min. Then, 60 μL of AAPH (14 mM) was added and the mixture was incubated for 80 min, reading the fluorescence at 485 and 520 nm of excitation and emission wavelengths, respectively, in a FLUOstar OPTIMA plate reader (BMG Labtech, Offenburg, Germany). The equipment was controlled by the FLUOstar Control ver. 1.32 R2 software. Final ORAC-FL value was expressed as μmol TE/mg of protein (mean of three replicates).

2.6. *Protective Effects in Macrophage RAW264.7*

2.6.1. Cell Culture

The mouse macrophage cell line RAW264.7 (American Type Culture Collection, ATCC, Rockville, MD, USA) was grown in DMEM medium supplemented with 10% of FBS (*v/v*) and 1% penicillin/streptomycin (*v/v*). Cells were seeded in 75 cm^2 culture flasks and incubated at 37 °C in a humidified incubator containing 5% CO_2 and 95% air. Culture medium was changed every 2 days and subcultures were made by scraping.

2.6.2. Effects on Cell Viability

Cell viability was determined using the MTT assay. RAW264.7 cells were seeded onto 96-well plates at a density of 6×10^4 cells/well in complete medium with 10% FBS and incubated for 24 h at 37 °C. After removing the culture medium, samples were added. In the case of lipopolysaccharide (LPS)-stimulated cells, 20 μL of LPS (100 ng/mL, final concentration) was also added. Once incubated for 24 h at 37 °C, the supernatant was removed, and cells were washed with PBS. Then, an MTT solution (5 mg/mL in PBS) was added and the plate was incubated for 120 min at 37 °C. Once the supernatant was aspirated, insoluble formazan crystals were dissolved in dimethyl sulfoxide (DMSO), and the absorbance was measured at 570 nm in the Multiskan FC plate reader (ThermoTM Scientific, Wilmington, DE, USA). The results were expressed as percentage of the control, considered as 100%.

2.6.3. Effects on ROS Generation

Intracellular ROS levels were quantified according to the method described by LeBel et al. [23] using dichlorofluorescein (DCFH-DA) as probe. RAW264.7 macrophages were seeded onto 48-well plates (4.75×10^4 cells/well) in complete medium with 10% FBS and incubated for 24 h at 37 °C. Once the medium was aspirated, samples were added and cells were incubated for 24 h at 37 °C. In stimulated cells, 20 μL of LPS (100 ng/mL) was also added. After aspirating the supernatant, 100 μL of a solution containing 5 mM DCFH-DA dissolved in Hank's balanced salt solution (HBSS, Sigma-Aldrich) was added to the wells, incubating the plate at 37 °C for 60 min. The fluorescence was measured in a FLUOstar OPTIMA plate reader (BMG Labtech) at $\lambda_{excitation}$ of 485 and $\lambda_{emission}$ of 530 nm. The results were expressed as ROS levels (% compared with the control, considered as 100%).

2.6.4. Effects on NO Levels

RAW264.7 macrophages were seeded onto 96-well plates (1×10^5 cells/well) in complete medium with 10% FBS and incubated for 24 h at 37 °C. Once the medium was aspirated, samples were added (100 μL) and cells were incubated for 24 h at 37 °C. A volume of 20 μL of LPS (100 ng/mL) was added to stimulated cells. Nitrite accumulation, and indicator of nitric oxide (NO) synthesis, was measured in the macrophage culture medium by the Griess reaction following a previously described method [24]. Briefly, a mixture containing 100 μL of supernatant and 100 μL of Griess reagent [1% (*w/v*) sulfanyl amide and 0.1% (*w/v*) N-1-(naphthyl) ethylenediamine-di-HCl in 2.5% (*v/v*) H_3PO_4] was incubated for 15 min and the absorbance measured at 550 nm in a Synergy HTX microplate reader (BioTek Instruments, Inc.). A sodium nitrite standard curve (3.125–100 μM) was used to measure the amount of NO. Three independent experiments were conducted and data were expressed as the mean and standard deviation (SD) (n = 12).

2.7. Statistical Analysis

All data were analyzed in three independent experiments, and results were expressed as the mean ± SD. Data were analyzed using one-way analysis of variance (ANOVA). All analyses were run with the program GraphPad Prism v.9.0.1 (GraphPad Software, San Diego, CA, USA).

3. Results and Discussion

3.1. Characterization of Chachafruto Seed Protein Concentrate (CPC): Behavior under Simulated Gastrointestinal Digestion

The moisture, lipid, ash, and fiber content of CPC was 9.78, 0.20, 3.64, and 1.42% (dry weight), respectively. The protein content of CPC was 82.5%, higher than the contents of 65.43 and 62.00% reported by Intiquilla et al. [5] and Arango et al. [25], respectively. This variation could be due to the different protein content present in the cultivars resulting from the growing conditions and soil types [26]. The extraction conditions could also contribute to the observed differences. Thus, alkaline extraction followed by acid precipitation used by Villafuerte et al. [27] resulted in higher protein content of the chachafruto isolate (88.99%), comparable to the value obtained in our study.

The composition of amino acids in CPC, expressed as g/100 g protein, is reported in Table 1. Seventeen amino acids were detected in the CPC, while tryptophan was not observed because of the analysis conditions that provoked the destruction of this amino acid by acid hydrolysis. Among the essential amino acids (EAA), lysine and leucine were the most abundant, with values of 4.00 ± 0.13 and 6.02 ± 0.17 g/100 g protein, respectively. Aspartic acid + asparagine and glutamic acid + glutamine were the most abundant among the non-essential amino acids (NEAA), with values of 8.49 ± 0.31 and 12.34 ± 0.36 g/100 g of protein, respectively. These values were higher than those reported by Intiquilla et al. [5] for the flour, concentrate and alcalase hydrolysate from *E. edulis*. Chachafruto amino acid profile has been reported to be similar to that reported for other legumes such as cowpea [28], peas and lentils [29], being deficient in sulfur amino acids, such as methionine and tryptophan. The EAA/TAA and EAA to NEAA ratios were 40% and 67%, respectively, similar to those proposed by FAO/WHO for the protein reference pattern (EAA/TAA: 40%; EAA/NEAA: 60%) [30]. CPC was subjected to gastrointestinal digestion simulating physiological conditions. The protein profile of CPC before and after the digestive process was evaluated by SDS-PAGE (Figure 1). This analysis revealed the presence of proteins with molecular weight ranged from 2 to 150 kDa in CPC (lane 2), with high intensity bands for 15–50 kDa proteins. The intensity of these bands decreased in GDCPC (lane 7) and GIDCPC (lane 11), indicating the partial susceptibility of chachafruto proteins to the action of digestive enzymes. Globulins and albumins are the major components of pulse proteins. Globulins are divided into two groups by their sedimentation coefficients, the 7 S fraction (vicilins) and 11 S fraction (legumins). The band ≈ 50 kDa might be assigned to vicilin, an oligomeric protein made of three polypeptide subunits (α, β, and γ) [31]. The polypeptides at 20 and ≤14 kDa visible in GDCPC and GIDCPC might correspond to fragments resulting from the action of digestive enzymes on vicilin [32].

Table 1. Amino acid composition (g/100 g protein and g/100 g product) of *Erythrina edulis* protein concentrate from seed (CPC).

Amino Acid	Amino Acid Content (g/100 g Protein)	Amino Acid Content (g/100 g Product)	FAO Recommendations
Essential			
Lysine	4.00 ± 0.13	3.30 ± 0.11	5.20
Tryptophan	n.d.	n.d.	0.70
Phenylalanine	3.38 ± 0.10	2.78 ± 0.08	4.60 [a]
Tyrosine	3.76 ± 0.08	3.10 ± 0.06	
Methionine	1.01 ± 0.08	0.83 ± 0.06	2.60 [b]
Cysteine	0.90 ± 0.14	0.74 ± 0.12	
Threonine	2.56 ± 0.06	2.11 ± 0.05	2.70
Leucine	6.02 ± 0.17	4.97 ± 0.14	6.30
Isoleucine	2.32 ± 0.08	1.91 ± 0.06	3.10
Valine	3.23 ± 0.02	2.67 ± 0.01	4.20
Non-essential			
Aspartic acid + asparagine	8.49 ± 0.31	7.00 ± 0.26	
Glutamic acid + glutamine	12.34 ± 0.36	10.18 ± 0.30	
Serine	4.69 ± 0.11	3.87 ± 0.09	
Histidine	1.68 ± 0.10	1.38 ± 0.08	
Arginine	3.52 ± 0.18	2.90 ± 0.15	
Alanine	2.80 ± 0.13	2.31 ± 0.11	
Proline	3.83 ± 0.09	3.16 ± 0.08	
Glycine	3.28 ± 0.15	2.70 ± 0.12	
TAA	67.81	55.91	
HAA	27.25	22.48	
AAA	7.13	5.89	

n.d. not determined; [a] phenylalanine + tyrosine; [b] methionine + cysteine; HAA: hydrophobic amino acids (alanine, valine, isoleucine, leucine, tyrosine, phenylalanine, tryptophan, methionine, proline, and cysteine); TAA: total amino acids; AAA: aromatic amino acids (phenylalanine, tryptophan, and tyrosine). Data are the mean of two determinations.

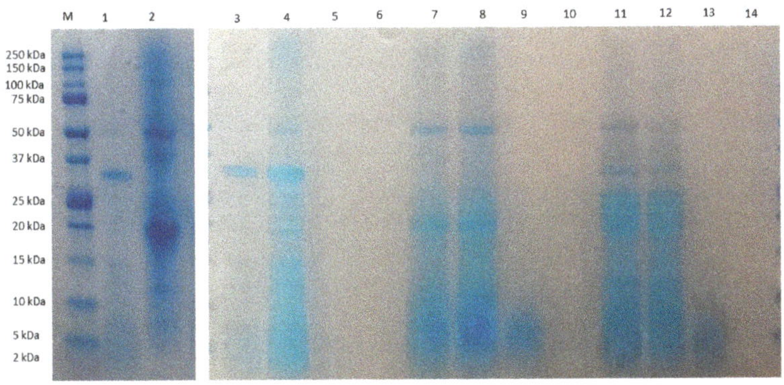

Figure 1. Electrophoretic (SDS-PAGE) analysis of *Erythrina edulis* (chachafruto) protein concentrate from seed (CPC) before and after its digestion simulating gastrointestinal conditions, and fractions collected from digested CPC by ultrafiltration. (M) molecular marker; (1,3) digestion blank (DB); (2) CPC; (4) fraction > 10 kDa from DB; (5) fraction 3–10 kDa from DB; (6) fraction < 3 kDa from DB; (7) gastric digest from CPC (GDCPC); (8) fraction > 10 kDa from GDCPC; (9) fraction 3–10 from GDCPC; (10) fraction < 3 kDa from GDCPC; (11) gastrointestinal digest from CPC (GIDCPC); (12) fraction > 10 kDa from GIDCPC; (13) fraction 3–10 kDa from GIDCPC; (14) fraction < 3 kDa from GIDCPC.

The protein profiles of the fractions higher than 10 kDa obtained from GDCPC (lane 8) and GIDCPC (lane 12) were similar to those of their corresponding complete digests (lanes 7 and 11). In the case of 3–10 kDa fractions, only slight bands corresponding to medium-weight peptides released during the digestive process were visible (lanes 9, 13). No bands were visible for the fractions lower than 3 kDa (lanes 10, 14), possibly due to the gel and electrophoretic conditions used, which only allowed detecting high molecular weight peptides. However, the peptide mass distribution of the digests determined by MALDI-TOF-MS confirmed that gastric and pancreatic enzymes digested the chachafruto proteins in small peptide fragments with overall sizes under 3500 Da: 75.8% and 87.7% of peptides found in GDCPC and GIDCPC, respectively had a molecular weight 500–1500 Da, while peptides of 1500–3500 Da represented 34.5% in GDCPC and 18.9% in GIDCPC (Figure 2).

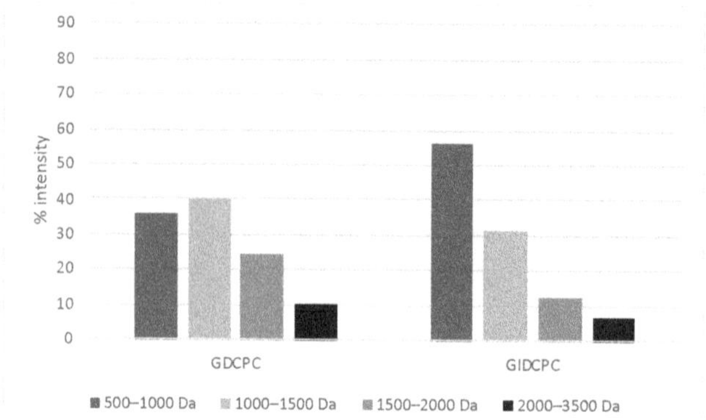

Figure 2. Peptide mass distribution (% intensity) of the gastric and gastrointestinal digests of chachafruto seeds protein concentrate (GDCPC and GIDCPC).

3.2. *Impact of the Gastrointestinal Digestion on the Antioxidant Activity of Chachafruto Seed Protein Concentrate (CPC)*

To evaluate the impact of the simulated digestive process on the release of bioactive peptides from chachafruto proteins, the antioxidant activity before and after digestion was studied by biochemical assays. At this time, the use of more than one method to study the antioxidant activity of a compound is suggested, evaluating its potentially mechanisms of action under different assay conditions [33]. In this study, the $ABTS^{\bullet+}$ and peroxyl ($ROO^{\bullet}$) radical scavenging activity of CPC, GDCPC, and GIDCPC and their ultrafiltered fractions were analyzed (Table 2).

Table 2. Antioxidant activity (ABTS and ORAC) of *Erythrina edulis* protein concentrate from seed (CPC), its gastric (GDCPC) and gastrointestinal (GIDCPC) digests and their fractions obtained by ultrafiltration.

Sample	TEAC (μmol TE/mg Protein)				ORAC (μmol TE/mg Protein)			
	Whole Sample	F > 10 kDa	F 3–10 kDa	F < 3 kDa	Whole Sample	F > 10 kDa	F 3–10 kDa	F < 3 kDa
CPC	0.10 ± 0.01	0.20 ± 0.00	0.47 ± 0.01	0.69 ± 0.07	0.47 ± 0.02	0.70 ± 0.01	0.79 ±0.01	4.29 ± 0.13
GDCPC	0.33 ± 0.02	0.17 ± 0.01	0.34 ± 0.00	0.66 ± 0.06	1.13 ± 0.02	0.47 ± 0.02	1.10 ± 0.02	1.93 ± 0.13
GIDCPC	0.46 ± 0.01	0.21 ± 0.01	0.30 ± 0.02	0.79 ± 0.00	1.12 ± 0.07	0.68 ± 0.02	0.96 ± 0.02	2.20 ± 0.18

TEAC: trolox equivalent antioxidant capacity; TE: trolox equivalent; ORAC: oxygen radical antioxidant capacity.

The ABTS assay measured the suppressive capacity of an antioxidant peptide against the radical $ABTS^{\bullet+}$. When added to medium containing $ABTS^{\bullet+}$, the peptides may act as electron or hydrogen donors, transforming this radical cation into the non-radical ABTS [34]. Although CPC showed slight capacity to scavenge the radical (TEAC value of 0.10 µmol TE/mg protein), the values notably increased after the action of gastric and mainly, pancreatic enzymes, reaching a TEAC value of 0.46 µmol TE/mg protein in GIDCPC. This value is within the range (0.35–0.91 µmol TE/mg protein) described by Intiquilla et al. [5] for CPC hydrolyzates by Neutrase® and Alcalase®. Same behavior was observed against the $ROO^{\bullet}$ radical, increasing the ORAC value up to 140% when chachafruto proteins were digested under gastrointestinal conditions. The ORAC assay is a broadly accepted method based on $ROO^{\bullet}$ scavenging for assessing food extracts that contain various antioxidants. It provides a single measurement of the inhibition time and degree. This assay uses the $ROO^{\bullet}$ radical as the reactant, with the redox potential and reaction mechanism similar to those of physiological oxidants. Another advantage of this method is the selection of physiological pH that allows the antioxidants reacting with an overall charge and protonation state similar to that occurring in the body [35]. The main contributors on the radical scavenging capacity observed in our study were low molecular weight peptides. Thus, the TEAC and ORAC values of the fraction F < 3 kDa were 0.79 µmol TE/mg protein and 2.20 µmol TE/mg protein, respectively (Table 2). These results agree with previous studies carried out with animal protein-derived fractions that demonstrated that peptides lower than 3 kDa showed the greatest antioxidant activity [36]. However, contradictory studies found the highest antioxidant activity in fractions containing high molecular weight peptides [37]. The highest activity of low molecular weight peptides might be due to their improved accessibility to the redox reaction system or the degree of freedom of the critical amino acid residues [38]. In addition to peptide length, amino acid composition is another contributor to the antioxidant activity of food-derived peptides. Thus, the presence of hydrophobic amino acids has been associated with the antioxidant properties of peptides [39,40]. In the CPC, HAA represent 40.2% of TAA (Table 1), thus, they could contribute on the antioxidant activity of released peptides after gastrointestinal digestion.

3.3. Effect of Chachafruto Seed Protein Concentrate (CPC) on Macrophage RAW264.7 Cells under Basal and Stimulated Conditions

A macrophage cell model was used to evaluate the antioxidant and immunomodulatory effects of chachafruto protein-derived peptides under basal and stimulated conditions. Firstly, the effect of CPC and its gastric and gastrointestinal digests on cell viability was evaluated after 24 h treatment at concentrations of 5–200 µg protein/mL (Figure 3). LPS (100 ng/mL), used as positive control, significantly reduced the cell viability up to 86.62 ± 6.53%. A significant and dose-dependent reduction was also observed when basal cells were treated with CPC (at doses higher than 10 µg protein/mL), indicating the potential cytotoxic effects of chachafruto proteins (Figure 3A). After hydrolysis by digestive enzymes, the cytotoxic effects of CPC were partially reduced, and cell viability inhibition was observed when cells were treated with CPC digests at doses higher than 50 µg protein/mL). Moreover, an induction of the cell viability of macrophages was observed when cells were treated with 5 µg protein/mL. This protective effect at low doses was also observed in LPS-challenged macrophages (Figure 3B). This result was consistent with previous studies that reported protective effects of natural bioactive compounds or food extracts against apoptosis induction by LPS through the modulation of different molecular pathways [41].

LPS has multiple effects on macrophages, including the induction of a variety of inflammatory modulators such as interleukin (IL)-1, tumor necrosis factor (TNF-α), and NO, among others [42]. A large amount of NO, particularly synthesized by iNOS, stimulates an inflammatory response to inhibit the growth of invading microorganisms and tumor cells [43]. Figure 4 summarizes the results obtained by exposing macrophages to CPC, GDCPC, GIDCPC and their fractions higher than 10 kDa. Under basal conditions, macrophages did not produce NO. However, challenging with LPS at 100 ng/mL resulted in a significant increase of released NO (49.83 μM ± 3.96), as has been previously reported [44]. Dose-dependent NO induction was also observed when basal cells were treated with CPC, reaching 80.35 μM ± 4.57 at 100 μg protein/mL ($p < 0.001$) (Figure 4A). Among ultrafiltered fractions, those containing medium and large peptides were the major contributors to the NO induction caused by CPC. Thus, at 100 μg protein/mL, the NO levels measured in cells treated with >10 kDa and 3–10 kDa fractions were 24.31 μM ± 2.11 (Figure 4A) and 39.13 μM ± 0.97, respectively. However, treatment of cells with a fraction lower than 3 kDa did not result in any increase in NO levels (data not shown) unlike that reported by Li and coworkers [45]. These authors found that the <3 kDa peptide in *Tricholoma matsutake* water extract induced NO production by macrophages at doses ranging from 100 to 400 μg/mL. In basal cells treated with GDCPC and GIDCPC, the increase in NO production was also dose-dependent, although the maximum reached at 100 μg protein/mL was lower than that determined in CPC-treated cells (Figure 4C,E). This result suggests that undigested proteins and polypeptides were responsible for the NO-inducing properties of CPC. In the digests, only fraction >10 kDa induced NO release, confirming the ability of high molecular weight peptides to exert an immunostimulant effect on macrophage cells.

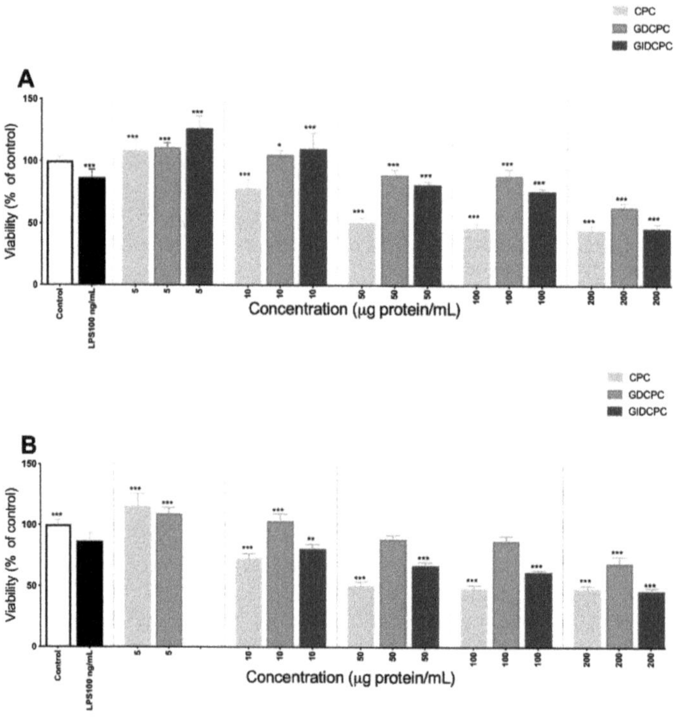

Figure 3. Viability (expressed as % of control considered as 100% cell viability) of (**A**) basal RAW264.7 macrophages and (**B**) lipopolysaccharide (LPS)-stimulated RAW264.7 macrophages, after 24 h incubation with different concentrations (5–200 μg protein/mL) of undigested and digested chachafruto seed protein concentrate (CPC, GDCPC, and GIDCPC). Significant differences compared to the control (* $p < 0.05$; ** $p < 0.01$; *** $p < 0.0001$).

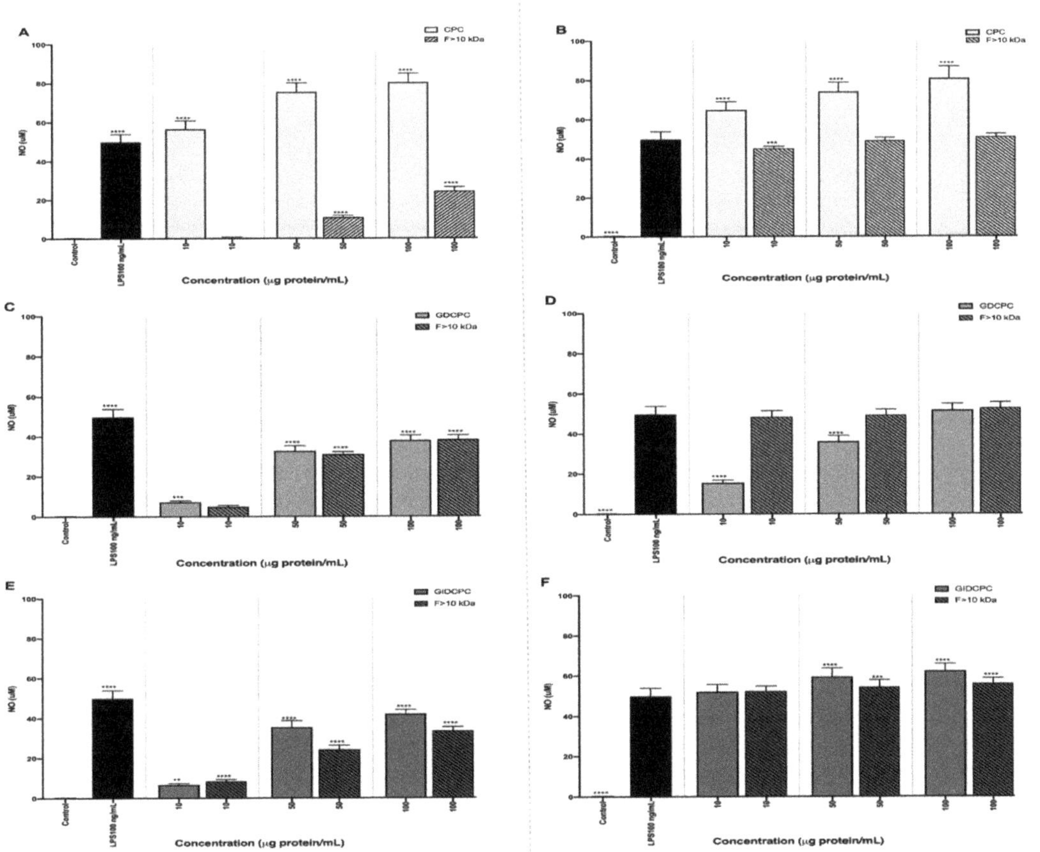

Figure 4. Nitric oxide (NO) production (µM) by (**A**,**C**,**E**) basal and (**B**,**D**,**F**) LPS-stimulated RAW264.7 macrophages after 24 h of exposure to (**A**,**B**) chachafruto protein concentrate (CPC) and its fraction higher than 10 kDa, (**C**,**D**) gastric digest from CPC (GDCPC) and its fraction higher than 10 kDa, (**E**,**F**) gastrointestinal digest from CPC (GIDCPC) and its fraction higher than 10 kDa. Significant differences compared to the control (** $p < 0.01$; *** $p < 0.001$; **** $p < 0.0001$).

In challenged cells, CPC potentiated the effect of LPS at a dose-dependent manner, reaching NO levels of 80.75 µM ± 6.05 (Figure 4B), similar to that determined in non-induced and CPC-treated cells. However, its fraction > 10 kDa at 10 µg protein/mL reduced the stimulating effect caused by LPS, without exerting any effect at higher doses. The reverting effect was also observed when macrophages were co-incubated with GDCPC and LPS, reducing NO levels up to 15.45 µM ± 1.35 when the gastric digest was used at 10 µg protein/mL (Figure 4D). No effects were observed on LPS-induced cells when GDCPC was used at doses higher than 50 µg protein/mL or when cells were co-incubated with LPS and its fraction > 10 kDa. However, both GIDCPC and its high molecular weight peptide fraction potentiated the NO-inducing effect caused by LPS (Figure 4F). These results indicate that peptides contained in GDCPC could protect cells from the LPS challenge while high molecular weight peptides contained in CPC and its gastrointestinal digest activated macrophages to release inflammatory mediators such as NO. This is a free radical capable of reacting with superoxide anion to generate a selective oxidant and nitrating agent, peroxynitrite, that interacts with biological molecules and damages the cell membranes, being able to result in cell death [46]. Although properly regulated inflammatory responses are necessary for healthy immunofunction, an excessive inflammation may cause a chronic inflammatory status, sepsis, and even death [47]. Numerous peptides from legumes are considered to potentially work as antioxidant and anti-inflammatory agents through quenching free radicals, increasing antioxidant defenses, or inhibiting the release of proinflammatory mediators such as NO. As an example, digests from lupine protein contained different peptides in the extruded lupin seed flour fraction 2 (ELPF-2), which significantly reduced the production of NO and other mediators like TNF-α, IL-1β in macrophages [48]. Immunomodulatory capacities of food-derived peptides have been reported to be affected by their amino acid composition, length, sequence, and molecule structure [49].

To evaluate the effects of chachafruto protein-derived peptides on the oxidative status of cells, ROS levels were measured using the DCFH-DA probe in RAW264.7 cells under basal conditions (Table 3). The evaluation of intracellular ROS content is considered a good indicator of the oxidative damage in living cells [50]. LPS (100 ng/mL), as positive control, induced ROS generation (186.12% ± 16.43) as previously reported [51]. CPC, GDCPC and its fraction >10 kDa exerted a prooxidant effect at high doses. However, antioxidant activity was observed when cells were treated with these samples at low doses (Table 3). Fractions containing medium and low molecular weight peptides exerted protective effects at all assayed doses, indicating that these peptides were the main responsible for the alleviating effects on the oxidative status. Similar protective effects against oxidative stress were reported for the low molecular weight peptides extracted from the edible fungi *T. matsutake* [45]. Our results were in agreement with the observed in vitro antioxidant activity. However, in LPS-challenged cells, the decrease in ROS production occurred only in CPC F < 3 kDa at 100 μg/mL and GIDCPC F 3–10 kDa at 50 μg/mL and F < 3 kDa at 10 μg/mL (data not shown). Peptide fractions and peptides isolated from legumes have been reported to possess antioxidant properties in cell models. Martínez-Leo et al. [52], assessed the antioxidant and protective effect of the peptide fractions derived from *Mucuna pruriens*. In this study, the highest antioxidant activity was exhibited by the 1–3 kDa fraction that inhibited ROS production in H_20_2-treated HeLa cells, lowering the intracellular levels by 207 ± 4.20%. Hernández-Ledesma et al. [53], demonstrated the capacity of soybean peptide lunasin to decrease ROS levels in LPS-activated macrophages at concentrations ranged from 50 μM to 200 μM. Also, the release of antioxidant peptides during simulated gastrointestinal digestion of red bean protein concentrate was reported by Piñuel et al. [54]. These authors demonstrated that after the action of gastric and pancreatic enzymes, peptides released to red bean proteins were able to reduce the intracellular ROS levels up to 75.30% and 66.40%, respectively.

Table 3. Effects of *Erythrina edulis* protein concentrate from seed (CPC), its gastric (GDCPC) and gastrointestinal (GIDCPC) digests, and their fractions obtained by ultrafiltration on reactive oxygen species (ROS) generation in macrophage RAW264.7 cells under basal conditions.

Sample	Concentration (μg/mL)	ROS Generation (% of Control)			
		Whole Sample	Fraction > 10 kDa	Fraction 3–10 kDa	Fraction < 3 kDa
Control	-	100.00 ± 4.87			
LPS	0.1	186.12 ± 16.43 ***			
CPC	5.0	65.55 ± 3.69 ***	85.73 ± 6.26 *	57.03 ± 4.40 ***	52.84 ± 4.43 ***
	10.0	102.56 ± 10.56	116.25 ± 6.47 *	58.14 ± 4.44 ***	50.54 ± 3.28 ***
	50.0	140.62 ± 8.44 ***	188.80 ± 8.90 ***	60.21 ± 3.85 ***	57.26 ± 5.61 ***
	100.0	163.20 ± 15.91 ***	225.02 ± 6.63 ***	66.48 ± 5.97 ***	61.19 ± 3.34 ***
GDCPC	5.0	41.96 ± 5.04 ***	70.56 ± 6.00 ***	61.70 ± 2.30 ***	58.80 ± 2.69 ***
	10.0	50.76 ± 6.12 ***	75.00 ± 5.17 ***	57.97 ± 2.28 ***	60.67 ± 2.97 ***
	50.0	115.53 ± 10.16 *	199.00 ± 19.47 ***	82.02 ± 5.62 ***	70.23 ± 3.21 ***
	100.0	137.26 ± 8.69 ***	236.38 ± 22.35 ***	118.59 ± 5.92 **	75.19 ± 4.69 ***

Table 3. *Cont.*

Sample	Concentration (μg/mL)	ROS Generation (% of Control)			
		Whole Sample	Fraction > 10 kDa	Fraction 3–10 kDa	Fraction < 3 kDa
GIDCPC	5.0	67.56 ± 3.77 ***	66.30 ± 4.86 ***	53.76 ± 4.38 ***	61.02 ± 2.51 ***
	10.0	74.00 ± 6.83 **	106.74 ± 10.26	55.42 ± 5.23 ***	66.99 ± 2.13 ***
	50.0	224.44 ± 19.08 ***	180.46 ± 14.09 ***	68.96 ± 3.13 ***	73.91 ± 4.23 ***
	100.0	306.17 ± 26.57 ***	245.40 ± 8.98 ***	70.23 ± 5.56 ***	77.81 ± 2.30 ***

Data indicate mean value ± standard deviation. Significant differences compared to the control (* $p < 0.01$; ** $p < 0.001$; *** $p < 0.0001$).

4. Conclusions

After the action of gastric and pancreatic enzymes, the protective effects of chachafruto proteins against oxidative stress were induced through the capacity shown by released low molecular weight peptides to scavenge radicals. Moreover, high molecular weight peptides contained in the CPC and GIDCPC were able to activate macrophages to release immunoresponse-associated mediators such as NO, while peptides contained in the GDCPC could protect cells from LPS challenge, reducing that biomarker. In conclusion, chachafruto proteins can be considered an alternative protein source with promising nutritional and biological properties after degradation in the gastrointestinal tract, releasing essential amino acids and bioactive peptides. Further research focused on the identification of peptides responsible for the beneficial effects and the confirmation of their bioavailability should be conducted to promote the use of these proteins as a new antioxidant and immunostimulatory ingredient in functional foods and nutraceuticals.

Author Contributions: Conceptualization, J.L.C., B.H.-L. and J.E.Z.; investigation, J.L.C.; writing–review and editing, J.L.C., B.H.-L. and J.E.Z. All authors contributed equally to this work. All authors have read and agreed to the published version of the manuscript.

Funding: This research was supported by MINCIENCIAS, through credit scholarship 755 "Doctorados nacionales 2016".

Institutional Review Board Statement: Not applicable.

Informed Consent Statement: Not applicable.

Data Availability Statement: Not applicable.

Conflicts of Interest: The authors declare no conflict of interest.

References

1. Vermeir, I.; Weijters, B.; De Houwer, J.; Geuens, M.; Slabbinck, H.; Spruyt, A.; Van Kerckhove, A.; Van Lippevelde, W.; De Steur, H.; Verbeke, W. Environmentally Sustainable Food Consumption: A Review and Research Agenda from a Goal-Directed Perspective. *Front. Psychol.* **2020**, *11*, 1603. [CrossRef] [PubMed]
2. Pihlanto, A.; Mattila, P.; Mäkinen, S.; Pajari, A.-M. Bioactivities of Alternative Protein Sources and Their Potential Health Benefits. *Food Funct.* **2017**, *8*, 3443–3458. [CrossRef] [PubMed]
3. Otero, D.M.; Mendes, G.D.L.; Lucas, A.J.D.; Christ-Ribeiro, A.; Ribeiro, C.D.F. Exploring Alternative Protein Sources: Evidence from Patents and Articles Focusing on Food Markets. *Food Chem.* **2022**, *394*, 133486. [CrossRef] [PubMed]
4. Sánchez Chero, M.J.; Sánchez Chero, J.A.; Miranda Zamora, W. Technify and Conserve the Bioactive Components of Pashul (*Erythrina edulis*) for Human Consumption. *UCV HACER Rev. Inv. Cult.* **2019**, *8*, 11–17.
5. Intiquilla, A.; Jiménez-Aliaga, K.; Zavaleta, A.I.; Arnao, I.; Peña, C.; Chávez-Hidalgo, E.L.; Hernández-Ledesma, B. *Erythrina edulis* (Pajuro) Seed Protein: A New Source of Antioxidant Peptides. *Nat. Prod. Commun.* **2016**, *11*, 781–786. [CrossRef] [PubMed]
6. Intiquilla, A.; Jiménez-Aliaga, K.; Guzmán, F.; Alvarez, C.A.; Zavaleta, A.I.; Izaguirre, V.; Hernández-Ledesma, B. Novel Antioxidant Peptides Obtained by Alcalase Hydrolysis of *Erythrina edulis* (Pajuro) Protein. *J. Sci. Food Agric.* **2019**, *99*, 2420–2427. [CrossRef]
7. Palma-Albino, C.; Intiquilla, A.; Jiménez-Aliaga, K.; Rodríguez-Arana, N.; Solano, E.; Flores, E.; Zavaleta, A.I.; Izaguirre, V.; Hernández-Ledesma, B. Albumin from *Erythrina edulis* (Pajuro) as a Promising Source of Multifunctional Peptides. *Antioxidants* **2021**, *10*, 1722. [CrossRef]

8. Vilcacundo, R.; Miralles, B.; Carrillo, W.; Hernández-Ledesma, B. In Vitro Chemopreventive Properties of Peptides Released from Quinoa (*Chenopodium quinoa* Willd.) Protein under Simulated Gastrointestinal Digestion. *Food Res. Int.* **2018**, *105*, 403–411. [CrossRef]
9. Phongthai, S.; D'Amico, S.; Schoenlechner, R.; Homthawornchoo, W.; Rawdkuen, S. Fractionation and Antioxidant Properties of Rice Bran Protein Hydrolysates Stimulated by In Vitro Gastrointestinal Digestion. *Food Chem.* **2018**, *240*, 156–164. [CrossRef]
10. Nascimento, E.S.D.; Anaya, K.; de Oliveira, J.M.C.; de Lacerda, J.T.J.G.; Miller, M.E.; Dias, M.; Mendes, M.A.; Pallone, J.d.A.L.; Arns, C.W.; Juliano, M.A.; et al. Identification of Bioactive Peptides Released from In Vitro Gastrointestinal Digestion of Yam Proteins (*Dioscorea cayennensis*). *Food Res. Int.* **2021**, *143*, 110–286.
11. Cipollone, M.A.; Tironi, V.A. Yellow Pea Flour and Protein Isolate as Sources of Antioxidant Peptides after Simulated Gastrointestinal Digestion. *Legume Sci.* **2020**, *2*, e59. [CrossRef]
12. Tagliazucchi, D.; Martini, S.; Bellesia, A.; Conte, A. Identification of ACE-Inhibitory Peptides from *Phaseolus vulgaris* after In Vitro Gastrointestinal Digestion. *Int. J. Food Sci. Nutr.* **2015**, *66*, 774–782. [CrossRef] [PubMed]
13. Luna-Vital, D.; de Mejía, E.G. Peptides from Legumes with Antigastrointestinal Cancer Potential: Current Evidence for Their Molecular Mechanisms. *Curr. Opin. Food Sci.* **2018**, *20*, 13–18. [CrossRef]
14. Wang, S.; Rao, P.; Ye, X. Isolation and Biochemical Characterization of a Novel Leguminous Defense Peptide with Antifungal and Antiproliferative Potency. *Appl. Microbiol. Biotechnol.* **2009**, *82*, 79–86. [CrossRef] [PubMed]
15. Pavlicevic, M.; Marmiroli, N.; Maestri, E. Immunomodulatory Peptides—A Promising Source for Novel Functional Food Production and Drug Discovery. *Peptides* **2022**, *148*, 170696. [CrossRef] [PubMed]
16. Chalamaiah, M.; Yu, W.; Wu, J. Immunomodulatory and Anticancer Protein Hydrolysates (Peptides) from Food Proteins: A Review. *Food Chem.* **2018**, *245*, 205–222. [CrossRef]
17. Kiewiet, M.B.; Faas, M.M.; De Vos, P. Immunomodulatory Protein Hydrolysates and Their Application. *Nutrients* **2018**, *10*, 904. [CrossRef]
18. Cunniff, P.; Association of Official Analytical Chemists (Eds.) *Official Methods of Analysis of AOAC International*; AOAC: Washington, DC, USA, 1999; ISBN 978-0-935584-54-7.
19. Minekus, M.; Alminger, M.; Alvito, P.; Ballance, S.; Bohn, T.; Bourlieu, C.; Carrière, F.; Boutrou, R.; Corredig, M.; Dupont, D. A Standardised Static In Vitro Digestion Method Suitable for Food–An International Consensus. *Food Funct.* **2014**, *5*, 1113–1124. [CrossRef]
20. Lozano-Ojalvo, D.; Molina, E.; López-Fandiño, R. Hydrolysates of Egg White Proteins Modulate T- and B-Cell Responses in Mitogen-Stimulated Murine Cells. *Food Funct.* **2016**, *7*, 1048–1056. [CrossRef]
21. Re, R.; Pellegrini, N.; Proteggente, A.; Pannala, A.; Yang, M.; Rice-Evans, C. Antioxidant Activity Applying an Improved ABTS Radical Cation Decolorization Assay. *Free Radic. Biol. Med.* **1999**, *26*, 1231–1237. [CrossRef]
22. Hernández-Ledesma, B.; Dávalos, A.; Bartolomé, B.; Amigo, L. Preparation of Antioxidant Enzymatic Hydrolysates from α-Lactalbumin and β-Lactoglobulin. Identification of Active Peptides by HPLC-MS/MS. *J. Agric. Food Chem.* **2005**, *53*, 588–593. [CrossRef] [PubMed]
23. LeBel, C.P.; Ischiropoulos, H.; Bondy, S.C. Evaluation of the Probe 2′,7′-Dichlorofluorescin as an Indicator of Reactive Oxygen Species Formation and Oxidative Stress. *Chem. Res. Toxicol.* **1992**, *5*, 227–231. [CrossRef] [PubMed]
24. Martinez-Villaluenga, C.; Dia, V.P.; Berhow, M.; Bringe, N.A.; Gonzalez de Mejia, E. Protein Hydrolysates from β-Conglycinin Enriched Soybean Genotypes Inhibit Lipid Accumulation and Inflammation In Vitro. *Mol. Nutr. Food Res.* **2009**, *53*, 1007–1018. [CrossRef] [PubMed]
25. Arango Bedoya, O.; Bolaños Patiño, V.; Ricaurte García, D.; Caicedo, M.; Guerrero, Y. Obtención de un extracto proteico a partir de harina de chachafruto (*Erythrina edulis*). *Univ. Salud* **2012**, *14*, 161–167.
26. Klupšaitė, D.; Juodeikienė, G. Legume: Composition, Protein Extraction and Functional Properties. A Review. *Chem. Technol.* **2015**, *66*, 5–12. [CrossRef]
27. Villafuerte, F.; Pérez, E.; Mahfoud, A.; Valero, Y.; Enríquez, M.; Yanez, K.; Manobanda, P. Characerization of Erythrina Edulis Triana and Obtaining Protein Isolate. In Proceedings of the CICABI, Ambato, Ecuador, 25–29 June 2018; p. 233.
28. Teka, T.A.; Retta, N.; Bultosa, G.; Admassu, H.; Astatkie, T. Protein Fractions, In Vitro Protein Digestibility and Amino Acid Composition of Select Cowpea Varieties Grown in Ethiopia. *Food Biosci.* **2020**, *36*, 100634. [CrossRef]
29. Pastor-Cavada, E.; Juan, R.; Pastor, J.E.; Alaiz, M.; Vioque, J. Protein and Amino Acid Composition of Select Wild Legume Species of Tribe *Fabeae*. *Food Chem.* **2014**, *163*, 97–102. [CrossRef]
30. Xiong, S.; Yao, X.; Li, A. Antioxidant Properties of Peptide from Cowpea Seed. *Int. J. Food Prop.* **2013**, *16*, 1245–1256. [CrossRef]
31. Heldt, H.-W.; Piechulla, B. Chapter 12—Products of Nitrogen Fixation and Nitrate Assimilation Are Deposited as Storage Proteins. In *Plant Biochemistry*, 5th ed.; Heldt, H.-W., Piechulla, B., Eds.; Academic Press: Rostock, Germany, 2021; pp. 299–305, ISBN 978-0-12-818631-2.
32. Junejo, S.A.; Ding, L.; Fu, X.; Xiong, W.; Zhang, B.; Huang, Q. Pea Cell Wall Integrity Controls the Starch and Protein Digestion Properties in the Infogest In Vitro Simulation. *Int. J. Biol. Macromol.* **2021**, *182*, 1200–1207. [CrossRef]
33. Gómez, A.; Gay, C.; Tironi, V.; Avanza, M.V. Structural and Antioxidant Properties of Cowpea Protein Hydrolysates. *Food Biosci.* **2021**, *41*, 101074. [CrossRef]
34. Zhan, Q.; Wang, Q.; Liu, Q.; Guo, Y.; Gong, F.; Hao, L.; Wu, H.; Dong, Z. The Antioxidant Activity of Protein Fractions from Sacha Inchi Seeds after a Simulated Gastrointestinal Digestion. *LWT* **2021**, *145*, 111356. [CrossRef]

35. Jodayree, S.; Smith, J.C.; Tsopmo, A. Use of Carbohydrase to Enhance Protein Extraction Efficiency and Antioxidative Properties of Oat Bran Protein Hydrolysates. *Food Res. Int.* **2012**, *46*, 69–75. [CrossRef]
36. You, L.J.; Zhao, M.M.; Regenstein, J.M.; Ren, J.Y. Purification and Identification of Antioxidative Peptides from Loach (*Misgurnus anguillicaudatus*) Protein Hydrolysate by Consecutive Chromatography and Electrospray Ionization-Mass Spectrometry. *Food Res. Int.* **2010**, *43*, 1167–1173. [CrossRef]
37. Feng, L.; Peng, F.; Wang, X.; Li, M.; Lei, H.; Xu, H. Identification and Characterization of Antioxidative Peptides Derived from Simulated In Vitro Gastrointestinal Digestion of Walnut Meal Proteins. *Food Res. Int.* **2019**, *116*, 518–526. [CrossRef] [PubMed]
38. Peighambardoust, S.H.; Karami, Z.; Pateiro, M.; Lorenzo, J.M. A Review on Health-Promoting, Biological, and Functional Aspects of Bioactive Peptides in Food Applications. *Biomolecules* **2021**, *11*, 631. [CrossRef]
39. Sannasimuthu, A.; Kumaresan, V.; Anilkumar, S.; Pasupuleti, M.; Ganesh, M.-R.; Mala, K.; Paray, B.A.; Al-Sadoon, M.K.; Albeshr, M.F.; Arockiaraj, J. Design and Characterization of a Novel Arthrospira Platensis Glutathione Oxido-Reductase-Derived Antioxidant Peptide GM15 and Its Potent Anti-Cancer Activity via Caspase-9 Mediated Apoptosis in Oral Cancer Cells. *Free Radic. Biol. Med.* **2019**, *135*, 198–209. [CrossRef]
40. Ngo, N.T.T.; Shahidi, F. Antioxidant Properties of Camelina (*Camelina sativa* (L.) Crantz) Protein Hydrolysates. *J. Food Bioact.* **2021**, *16*, 75–82. [CrossRef]
41. Afrin, S.; Gasparrini, M.; Forbes-Hernández, T.Y.; Cianciosi, D.; Reboredo-Rodriguez, P.; Manna, P.P.; Battino, M.; Giampieri, F. Protective Effects of Manuka Honey on LPS-Treated RAW 264.7 Macrophages. Part 1: Enhancement of Cellular Viability, Regulation of Cellular Apoptosis and Improvement of Mitochondrial Functionality. *Food Chem. Toxicol.* **2018**, *121*, 203–213. [CrossRef]
42. Jeon, Y.J.; Han, S.H.; Lee, Y.W.; Lee, M.; Yang, K.H.; Kim, H.M. Dexamethasone Inhibits IL-1β Gene Expression in LPS-Stimulated RAW 264.7 Cells by Blocking NF-KB/Rel and AP-1 Activation. *Immunopharmacology* **2000**, *48*, 173–183. [CrossRef]
43. Kim, Y.; Moon, J.S.; Lee, K.S.; Park, S.Y.; Cheong, J.; Kang, H.S.; Lee, H.Y.; Do Kim, H. Ca^{2+}/Calmodulin-Dependent Protein Phosphatase Calcineurin Mediates the Expression of INOS through IKK and NF-KB Activity in LPS-Stimulated Mouse Peritoneal Macrophages and RAW 264.7 Cells. *Biochem. Biophys. Res. Commun.* **2004**, *314*, 695–703. [CrossRef]
44. Song, S.; Li, T.; Cheng, X.; Ji, X.; Gao, D.; Du, M.; Jiang, N.; Liud, X.; Mao, X. Sea Cucumber Peptides Exert Anti-Inflammatory Activity Through Suppressing NF-κB and MAPK and Inducing HO-1 in RAW264.7 Macrophages. *Food Funct.* **2016**, *7*, 2773. [CrossRef] [PubMed]
45. Li, M.; Dong, L.; Du, H.; Bao, Z.; Lin, S. Potential Mechanisms Underlying the Protective Effects of *Tricholoma matsutake* Singer Peptides Against LPS-Induced Inflammation in RAW264.7 Macrophages. *Food Chem.* **2021**, *353*, 129452. [CrossRef] [PubMed]
46. Kim, I.D.; Ha, B.J. Paeoniflorin Protects RAW 264.7 Macrophages from LPS-Induced Cytotoxicity and Genotoxicity. *Toxicol. In Vitro* **2009**, *23*, 1014–1019. [CrossRef]
47. Chao, W.-W.; Chung, Y.-C.; Shih, I.-P.; Wang, H.-Y.; Chou, S.-T.; Hsu, C.-K. Red Bean Extract Inhibits Lipopolysaccharide-Induced Inflammation and H_2O_2-Induced Oxidative Stress in RAW 264.7 Macrophages. *J. Med. Food* **2015**, *18*, 724–730. [CrossRef] [PubMed]
48. Gao, Y.; Zhang, X.; Ren, G.; Wu, C.; Qin, P.; Yao, Y. Peptides from Extruded Lupin (*Lupinus albus* L.) Regulate Inflammatory Activity via the P38 MAPK Signal Transduction Pathway in RAW 264.7 Cells. *J. Agric. Food Chem.* **2020**, *68*, 11702–11709. [CrossRef] [PubMed]
49. He, K.; Zeng, Y.; Tian, H.; Zhang, Z.; Zhang, H.; Huang, F.; Yu, F. Macrophage Immunomodulatory Effects of Low Molecular Weight Peptides from *Mytilus coruscus* via NF-κB/MAPK Signaling Pathways. *J. Funct. Foods* **2021**, *83*, 104562. [CrossRef]
50. Frassinetti, S.; Gabriele, M.; Caltavuturo, L.; Longo, V.; Pucci, L. Antimutagenic and Antioxidant Activity of a Selected Lectin-Free Common Bean (*Phaseolus vulgaris* L.) in Two Cell-Based Models. *Plant Foods Hum. Nutr.* **2015**, *70*, 35–41. [CrossRef] [PubMed]
51. Nishanth, R.P.; Jyotsna, R.G.; Schlager, J.J.; Hussain, S.M.; Reddanna, P. Inflammatory Responses of RAW 264.7 Macrophages upon Exposure to Nanoparticles: Role of ROS-NFκB Signaling Pathway. *Nanotoxicology* **2011**, *5*, 502–516. [CrossRef]
52. Martínez-Leo, E.E.; Martín-Ortega, A.M.; Acevedo-Fernández, J.J.; Moo-Puc, R.; Segura-Campos, M.R. Peptides from *Mucuna pruriens* L., with Protection and Antioxidant In Vitro Effect on HeLa Cell Line. *J. Sci. Food Agric.* **2019**, *99*, 4167–4173. [CrossRef] [PubMed]
53. Hernández-Ledesma, B.; Hsieh, C.-C.; de Lumen, B.O. Antioxidant and Anti-Inflammatory Properties of Cancer Preventive Peptide Lunasin in RAW 264.7 Macrophages. *Biochem. Biophys. Res. Commun.* **2009**, *390*, 803–808. [CrossRef]
54. Piñuel, M.L.; Vilcacundo, E.; Boeri, P.A.; Barrio, D.A.; Morales, D.; Pinto, A.; Moran, R.; Samaniego, I.; Carrillo, W. Extraction of Protein Concentrate from Red Bean (*Phaseolus vulgaris* L.): Antioxidant Activity and Inhibition of Lipid Peroxidation. *J. Appl. Pharm. Sci.* **2019**, *9*, 45–58.

nutrients

MDPI

Article

Peptidomic Characterization and Amino Acid Availability after Intake of Casein vs. a Casein Hydrolysate in a Pig Model

Pablo Jiménez-Barrios [1], Laura Sánchez-Rivera [1], Daniel Martínez-Maqueda [1], Yann Le Gouar [2], Didier Dupont [2], Beatriz Miralles [1] and Isidra Recio [1,*]

1 Instituto de Investigación en Ciencias de la Alimentación, CIAL (CSIC-UAM, CEI UAM+CSIC), Nicolás Cabrera, 9, 28049 Madrid, Spain
2 Institut National de Recherche Pour l'Agriculture, l'Alimentation et l'Environnement (INRAE) STLO Agrocampus Ouest, 35042 Rennes, France
* Correspondence: i.recio@csic.es; Tel.: +34-910017940; Fax: +34-910017905

Abstract: It is known that casein hydrolysis accelerates gastrointestinal transit in comparison to intact casein, although the effect of the protein hydrolysis on the composition of the digests is not fully understood. The aim of this work is to characterize, at the peptidome level, duodenal digests from pigs, as a model of human digestion, fed with micellar casein and a previously described casein hydrolysate. In addition, in parallel experiments, plasma amino acid levels were quantified. A slower transit of nitrogen to the duodenum was found when the animals received micellar casein. Duodenal digests from casein contained a wider range of peptide sizes and a higher number of peptides above five amino acids long in comparison with the digests from the hydrolysate. The peptide profile was markedly different, and although β-casomorphin-7 precursors were also found in hydrolysate samples, other opioid sequences were more abundant in the casein digests. Within the same substrate, the evolution of the peptide pattern at different time points showed minimal changes, suggesting that the protein degradation rate relies more on the gastrointestinal location than on digestion time. Higher plasma concentrations of methionine, valine, lysine and amino acid metabolites were found in animals fed with the hydrolysate at short times (<200 min). The duodenal peptide profiles were evaluated with discriminant analysis tools specific for peptidomics to identify sequence differences between both substrates that can be used for future human physiological and metabolic studies.

Keywords: casein; hydrolysate; duodenal digest; peptidomics; digestion kinetics

1. Introduction

Milk is an excellent source of highly digestible proteins in human nutrition. It is well known that the speed of protein digestion and amino acid absorption is affected by the protein nature and the technological treatment applied, and protein digestion kinetics modulate the metabolic response. While the casein fraction of milk, so-called slow protein, coagulates in the stomach and promotes postprandial protein deposition, whey proteins induce a fast increase in plasma amino acids, protein synthesis, and oxidation [1]. Small changes in processing have also been found to affect the kinetics of protein digestion. For instance, dairy gels differing in the mode of coagulation (acidification vs. renneting) cause different compositions of porcine duodenal effluents and plasma amino acids [2].

Enzymatic hydrolysis is increasingly used in the dairy sector for the design of ingredients for infant formulas [3]. In addition, casein hydrolysates have demonstrated to stimulate insulin secretion and to show glycemic control properties in both animals and humans [4,5]. A diet containing hydrolyzed casein has been reported to enhance upper gastrointestinal transit and to increase ileal microbial carbohydrate metabolism in comparison with an intact casein-based diet [6]. However, the influence of the enzymatic hydrolysis of the casein fraction of milk on gastric emptying and composition of the gastrointestinal contents is not fully understood. A previous study with aged rats found that casein or whey hydrolysates do not alter gastrointestinal transit compared with their respective intact counterparts. The slowed gastrointestinal transit of casein was reversed with opioid inhibitors, suggesting the involvement of opioid receptors [7]. Similarly, protein hydrolysis has been described to reduce the gastrointestinal transit time in preterm infants [8]. In rats, this effect was also attributed to the reduction of casein-derived opioid receptor agonists [9].

The advances in peptidomic techniques based on mass spectrometry (MS) have allowed for study of the complex protein digestome at various regions of the gastrointestinal tract [10,11]. These techniques have widely been used to characterize the large variety of peptides generated by the gastrointestinal proteases and peptidases, mainly in vitro [12–15]. However, in recent years, the characterization of gastrointestinal effluents from human or pig, as an accepted model of human digestion, is building a valuable peptide repertoire, especially after oral administration of milk proteins, to be used in physiological and metabolic studies or to validate in vitro digestion protocols [16–18].

The aim of this work is to characterize the peptidome of duodenal digests obtained in cannulated pigs after the oral administration of micellar casein and a casein hydrolysate. Different time points from 5 to 150 min were considered to evaluate kinetics of protein digestion in the duodenum. In addition, in parallel experiments, quantitative analysis of plasma amino acids was performed to assess the impact of protein hydrolysis on amino acid availability. Because pig is recognized as an excellent model of the human gastrointestinal tract, the results offer new information about the faster digestion and quicker availability of amino acids of protein hydrolysates used in human nutrition.

2. Materials and Methods

2.1. Animal Experiment

All procedures were in accordance with the European Community guidelines for the use of laboratory animals (L358-86/609/EEC). The local committee for ethics in animal experimentation approved the study. The animals were 6 Large White × Landrace × Piétrain pigs of about 40 kg live weight. Two weeks before starting the trials, they were surgically fitted with a T-shaped cannula in the duodenum (10 cm downstream from the pylorus). Catheterization of the external jugular vein was also performed. Following the operation, the animals were housed in individual slatted pens within a ventilated room with controlled temperature (21 °C). During convalescence following the surgical procedure and also between the sampling days, pigs were fed with 800 g/d of a pig feed concentrate (Cooperl Arc Atlantique®, Pelstan, France) containing 16% proteins, 1% fat, 4% cellulose and 5% mineral matter and had free access to water.

The test meals were commercial micellar casein (Prodiet 85B, batch no. 141179) purchased from Ingredia S. A., Arras, France, and a pepsin hydrolysate from this casein. The hydrolysate was prepared with food grade pepsin (P389P/3 Biocatalyst 3000 U/mg) at 2% enzyme/substrate ratio (*w*/*w*), added at 0 and 3 h during a hydrolysis time of 6 h at 37 °C, as previously reported [19]. The nitrogen content of casein and the hydrolysate was 77.8 and 77.5%, respectively, as determined by Kjeldahl.

The collection of duodenal effluents and blood samples was performed in separate assays. In each one, both dairy products were randomly tested on each animal. Within a period (2 weeks), the days of sampling were separated by at least 2 day. Test meals, casein and the casein hydrolysate (129 g of powder, on the basis of protein content) were reconstituted in water (250 mL) and offered to overnight fasted pigs for 10 min. The pigs had no access to water until 4 h after the meal delivery. Duodenum effluents were collected in plastic bottles 15 min before and 5, 10, 15, 20, 30, 45, 60, 90, 120, and 150 min after ingestion of test meal. The sampling was stopped when 40 mL was collected or after a maximum of 3 min of sampling time. A protease inhibitor (Pefabloc® at 2.5 mM final concentration) was previously added to the sample tubes. Collected effluents were weighed, freeze-dried and stored at −20 °C.

Blood sampling was performed at 15 min before and 5, 20, 45 min, 1, 2, 3, 4 and 6 h after ingestion of the casein or its hydrolysate. Blood samples (5 mL) were collected in heparinized-lithium tubes where a protease inhibitor cocktail for plasma samples was previously added (1 tablet/10 mL blood) (Complete Mini, EDTA-free, Roche Farma, Madrid, Spain). Then, centrifugation at 3000× g 4 °C for 10 min was performed, and the supernatant was frozen with liquid nitrogen and stored at −80 °C. Amino acids and related metabolites in plasma were determined by ion-exchange chromatography with ninhydrine at the Diagnosis of Molecular Diseases Centre from the Autonomous University of Madrid. Briefly, 2 mL of plasma was deproteinized with 200 µL 10% trichloroacetic acid. After incubation for 15 min on ice, the samples were centrifuged for 10 min at 1500× g and 4 °C, and 150 µL of supernatant was collected. The samples were placed in an autosampler protected from light, and 20 µL was injected.

2.2. Nitrogen Analysis and SDS-PAGE

Nitrogen content of duodenal effluents was determined by elemental analysis in a LECO chns-932 analyzer at the Elemental analysis unit from the Interdepartmental Investigation Service from the Autonomous University of Madrid. Freeze-dried digests were adjusted to 1 mg of protein/mL of sample buffer (tris-HCL 0.05 M pH 6.8, SDS 1.6% *w:v*, glycerol 8% *v:v*, bromophenol blue 0.002% *w:v*, and β-mercaptoethanol 2% *v:v*) based on the nitrogen content, and the SDS-PAGE was conducted in Precast Criterion XT 12% Bis-Tris gels (Bio-Rad, Richmond, CA, USA) using the running buffer (19:1, *v/v* water/XT MES 20×; Bio-Rad Laboratories, Richmond, CA, USA). A molecular weight marker (Precision Plus ProteinTM Unstained standard, Bio-Rad Laboratories, Richmond, CA, USA) was used. A prior running at 100 V for 5 min was performed, and then, samples were run at 150 V for 45 min. Gel was rinsed with water, then stained with Coomasie Blue (Instant blue; Expedeon, Swavesey, UK) for 30 min, and rinsed again with water for 10 min.

2.3. Peptidomic Analysis

Tandem mass spectrometry (HPLC-MS/MS) was conducted on an Agilent 1100 HPLC system (Agilent Technologies, Waldbron, Germany) coupled to an Esquire 3000 ion trap mass spectrometer (Bruker Daltonics, Bremen, Germany) as previously described [20]. A linear gradient from 0 to 45% solvent B (100:0.1, *v/v* acetonitrile:formic acid) for 120 min on a Mediterranean Sea C18 column, 150 × 2.1 mm (Teknokroma, Barcelona, Spain) was used at 0.2 mL/min. The freeze-dried digests were reconstituted in solvent A (100:0.1, *v/v* water:formic acid) and the supernatants after centrifugation at 13,000× g, for 10 min were analyzed. Spectra were recorded over the mass/charge (*m/z*) range 100–1200, and the target mass was set at 900 m/z. Data processing was conducted using the software Data Analysis, version 4.0 and Biotools, version 3.2 from Bruker Daltonics. MASCOT V2.4 (MatrixScience) was used for peptide identification with the use of homemade databases of major genetic variants of bovine caseins and porcine endogenous proteins. The parameters selected were: no enzyme, no post-translational modifications, and peptide tolerance 0.1% and 0.5 Da for MS and MS/MS, respectively. A manual revision for each identification was performed, regardless of the p value.

To carry out most peptidomic analysis steps, R scripts were used. R scripts are some lines of code using R programing language. Peptide size distribution of all samples was calculated. Peptide sequences were divided by substrate and then subdivided by size. Raw numbers and their percentages were shown using the R package "ggplot2" version 3. 4.0 [21]. Peptides previously grouped by substrate were analyzed using the R package "ggseqlogo" version 0.1 to obtain the SeqLogo figure [22]. The sequence logo shows the probability to find a particular amino acid at the *N*-terminal (1, 2, 3) or C-terminal (−3, −2, −1) end of the sequence. In the case of five-length peptides, the amino acid in the 3rd position at both terminal ends was the same.

2.4. Statistical Analysis

Principal components analysis (PCA) was performed using the amino acids within the identified peptides as variables. For each amino acid and each time point, the value was calculated as the sum of intensities of all overlapping peptides. A hierarchical clustering tree of samples using the intensity of the peptides was built by PermutMatrix software version 1.9.3.0 [23]. Peptides present in at least two time points were used. Blood samples were analyzed using GraphPad Prism software version 8.0.2. Differences between substrates over time were found through a two-way ANOVA, followed by the Bonferroni test. The results were graphed with the R package "ggplot2"version 3.4.0 [21].

3. Results and Discussion

Duodenal contents after oral administration of casein and its hydrolysate were collected and freeze-dried until analysis, and the nitrogen content of the dried effluents is shown in Figure 1. Duodenal digests from casein had higher nitrogen content than those from the hydrolysate at 90 and 150 min, showing its emptying over a more extended period, compared to the hydrolysate. Although at short time points (<30 min), hydrolysate effluents showed higher protein content than casein ones, due to the high variability between animals, differences did not reach significance. Because the effluents were immediately frozen in liquid nitrogen and freeze-dried, flow rates of liquid effluents could not be measured, and differences in the nitrogen content between substrates could be diminished. Previous reports in aged rats have shown unaltered gastrointestinal transit with hydrolyzed casein compared with its intact counterpart [7]. Casein was emptied from the stomach mainly in the form of peptides, although in some time points (15, 45, 60 min), slight bands of intact casein fractions were visible by SDS-PAGE (Figure S1, Supplementary Material). In addition, because the casein substrate contained a small amount of whey proteins (β-lactoglobulin and α-lactalbumin), bands corresponding to these proteins were also detected up to 30 min after oral administration. Due to their high solubility, whey proteins are rapidly emptied from the stomach [1], and the net disappearance of β-lactoglobulin from human upper jejunum was reported to be faster than for casein [24]. Similarly, Barbé and co-workers found β-lactoglobulin in porcine duodenum after the administration of raw liquid milk [25]. In contrast to casein, the electrophoretic protein profile of the porcine duodenal effluents after the hydrolysate administration slightly varied with time, and all protein material detected was below 10 kDa.

The peptide fraction of the duodenal effluents was characterized by HPLC-MS/MS. A total of 3863 milk-derived peptides, around 400 peptides per time point, were identified in porcine duodenum after the administration of casein, and a lower number, ca. 300 per time point, was found after the administration of the hydrolysate (Table 1). This lower number in the case of the hydrolysate can be due to the limitation of our mass spectrometry conditions to identify peptides shorter than five amino acids. In addition, Table 1 shows the number of common peptides found in all animals ($n = 6$) to illustrate the inter-individual variability found. When using casein as feeding, the peptide profile showed a large variety of peptide sizes where ca. 25% were above 10 amino acids long. However, as expected, the peptides longer than 10 amino acids were below 10% when the protein was administered in the form of hydrolysate (Figure 2). This peptide size distribution matches with the information provided by SDS-PAGE (Figure S1, Supplementary Material). The most abundant peptides identified in both substrates corresponded to peptides between six and nine amino acids, but this can be caused by the selectivity of the MS method employed to detect peptides within this size range. Despite this, the peptide profile found in the duodenum after the administration of both substrates was notably different (Figure 3A,B). After casein administration, the most abundant β-casein-derived peptides belonged to the C-terminal domain of the protein, i.e., 193–209, while peptides in regions 58–80, and 115–126 from β-casein are specifically overrepresented in the duodenal effluents from the hydrolysate (Figure 3A). Other authors have also reported the resistance of the C-terminal region of β-casein to gastric pepsin both in simulated gastric digestions [26,27] and in pig duodenum [18]. The slower gastrointestinal transit of casein in comparison with soy has been attributed to the involvement of opioid peptides since it was reversed with opioid inhibitors [7]. Although some studies have found an accelerated gastrointestinal transport with hydrolyzed protein infant formula due to the reduction of opioid agonists [8,9], the presence of opioid sequences depends on the hydrolysis conditions (enzyme specificity, time . . .). Under our hydrolysis conditions, the regions corresponding to β-casein opioid peptides, i.e., β-casomorphin-7 (^{60}YPFPGPI^{66}) [28] and neocasomorphin-6 (^{114}YPVEPF^{119}) [29] were preserved, and these peptides or precursors thereof comprised highly overrepresented regions in the duodenal effluents from the hydrolysate. For the other main casein fraction, α_{s1}-casein, differences in the duodenal peptide profile between substrates were less marked (Figure 3B), but the hydrolysate effluents showed a higher abundancy of regions 24–33, 166–172, and 180–196. The domains corresponding to previously reported opioid peptides (α-casomorphins), i.e., α_{s1}-casein ^{90}RYLGYLE^{96} [30] and ^{144}YFYPEL^{149} [31], were more abundant in casein duodenal effluents than in hydrolysate ones. Other α_{s1}-casein abundant domains, such as 24–33, were common for both substrates. The compared peptide profiles for α_{s2}-casein and κ-casein are shown in Figure S2A,B, where clear differences between both substrates were also found.

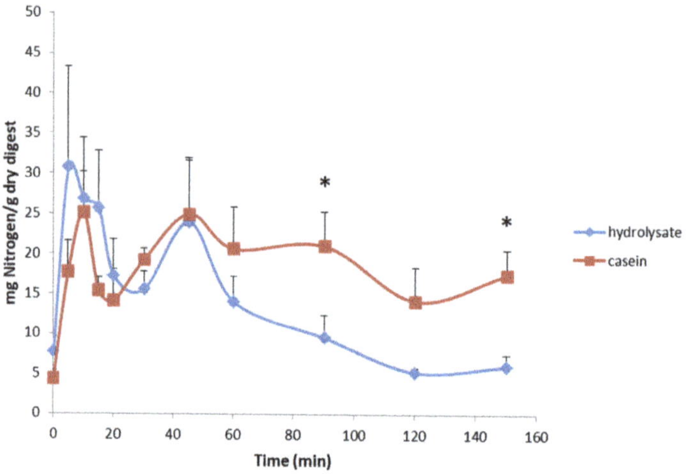

Figure 1. Average nitrogen content of pig duodenal digests (n = 6) over collection times at basal (−15 min), 5, 10, 15, 20, 30, 45, 60, 90, 120, and 150 min after intake of casein or hydrolysate. * indicates statistical difference ($p < 0.05$) by t test mean comparison.

In order to illustrate the evolution of the peptide profile with time, peptides were also aligned within the sequence of the main casein fractions, β-casein and α_{s1}-casein, and represented in the form of heatmaps (Figure S3A–D, Supplementary material). These graphs are based in the appearance frequency of each amino acid without taking into account the peptide intensity. Again, the C-terminal region of β-casein corresponded to the domain where a higher number of peptides was identified after casein administration, but for hydrolysate effluents, regions 70–100 and 115–125 were especially rich in peptides. Within the same substrate, differences in the peptide profile with time were less relevant, concluding that protein digestion occurs along the gastrointestinal tract, and for a given part, protein degradation is similar. Similar results were previously reported in human jejunum [16,17] or in different regions of the porcine gastrointestinal tract [18]. In addition to milk-derived peptides, the HPLC-MS/MS analysis of the effluents allowed us to identify several peptides from endogenous proteins (Table S1, Supplementary Material). The main identified proteins corresponded to endogenous enzymes: pepsin, chymotrypsin, carboxypeptidases, dipeptidyl peptidase or lipases, but also other peptides belonged to the sequence of gastrointestinal hormones, mucins or ubiquitous proteins, such as albumin.

Table 1. Total number of unique peptides derived from α_{s1}-, α_{s2}-, β-, and κ-casein identified in the duodenal effluents at each time point after intake of casein or hydrolysate, and those common among the assayed animals (n = 6).

Time	Casein		Hydrolysate	
	Total Peptide Number	Common Peptide Number [1]	Total Peptide Number	Common Peptide Number [1]
5	444	270	258	98
10	369	170	294	115
15	312	140	237	91
20	419	211	281	137
30	303	145	299	140
45	312	148	274	137
60	418	221	283	142
90	427	187	290	136
120	425	222	338	158
150	434	217	258	137
Total [2]	3863	1931	2812	1291

[1] In all animals (n = 6). [2] Total number of peptides and common peptide number among all time points.

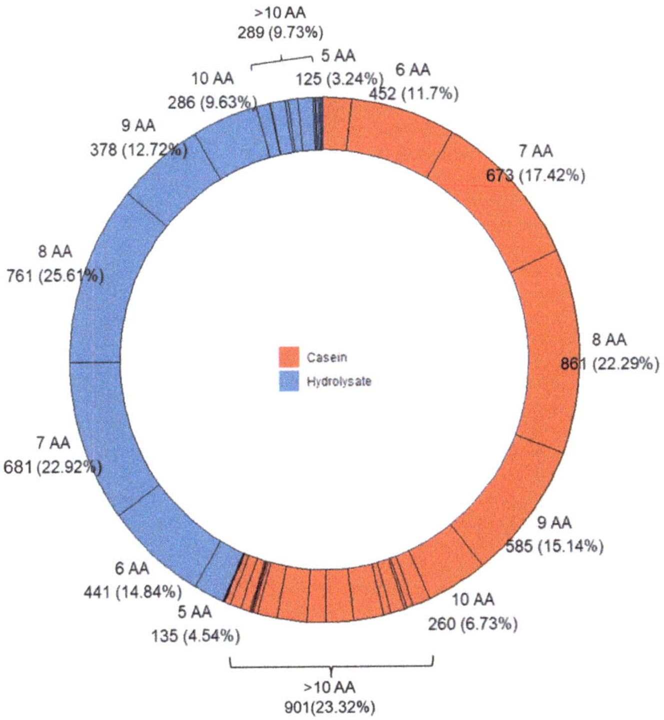

Figure 2. Distribution by peptide size of numbers of identified peptides derived from α_{s1}-, α_{s2}-, β-, and κ-casein in pig duodenal digests (n = 6) at basal (−15 min), 5, 10, 15, 20, 30, 45, 60, 90, 120, and 150 min after intake of casein or hydrolysate.

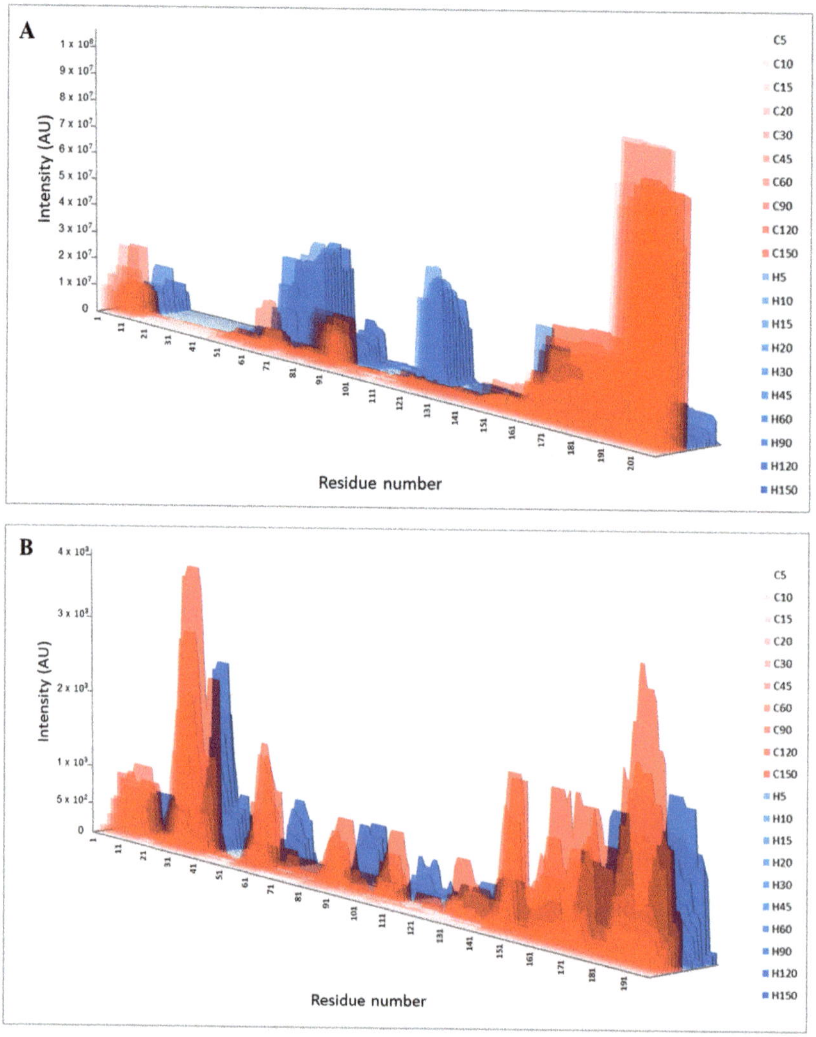

Figure 3. Identified peptides in pig duodenal digests (n = 6) after intake of casein (C) or hydrolysate (H) over collection time represented by area graphs. The height (Y-axis) is proportional to the sum of the identified peptide intensities. Profile peptides from β-casein (**A**) and α_{s1}-casein (**B**).

To compare the peptide composition of the effluents for both substrates at different time points in a holistic way, different techniques of discriminant analysis for peptidomics were used. Figure 4 shows the principal component analysis built with the peptide intensity values used in Figures 3 and S3 (including β-, α_{s1}-, α_{s2}, and κ-casein). The first component explained 41.6% of the variance while PC2 reached 22.1%. The peptide intensity profiles at the different time points were clearly grouped per substrate, demonstrating the differences shown in Figure 3 and the Supplementary Material. In addition, effluents obtained at longer time points, i.e., 90, 120 and 150 min, were separated from those at shorter times, and this effect was observed in both substrates, although it was more notable for casein duodenal effluents.

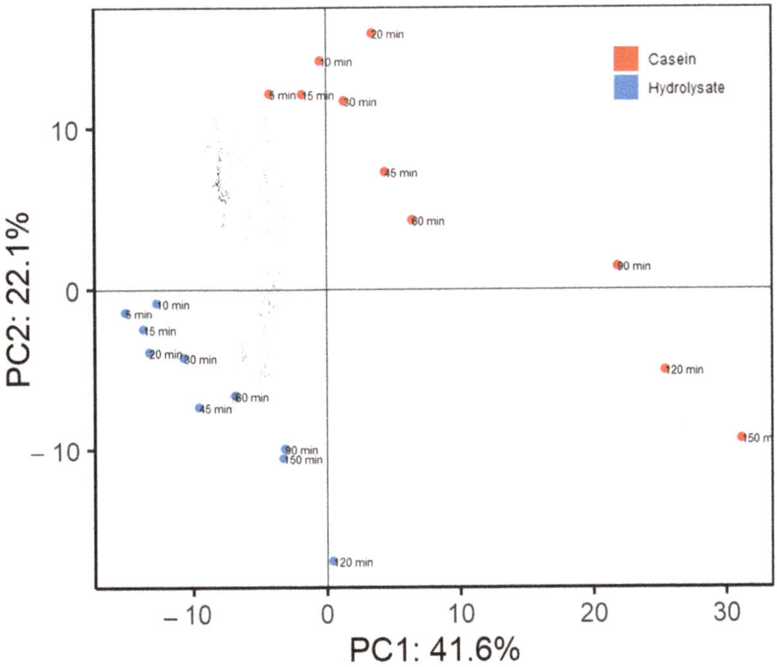

Figure 4. PCA analysis using, at each time point, the sum of intensities of all overlapping peptides identified in pig duodenal digests (n = 6) using each comprised amino acid as a variable.

The free software PermutMatrix was used to perform a graphical analysis of the peptide profiles based on sequence and intensity. The dendrogram in Figure 5 shows the clustering of samples per substrate, casein and hydrolysate. In addition, this analysis allows for the identification of the peptide differences in those regions overrepresented in each substrate. For instance, β-casomorphin precursors with valine or tyrosine at the *N*-terminal were highly detected in the effluents from the hydrolysate, but specific sequences, such as ^{57}SLVYPFPGPI^{66} or ^{58}LVYPFPGPI^{66}, were more abundant in casein samples (Figure 5B). Similarly, β-casein regions containing the antihypertensive tri-peptides IPP and VPP [32] were overrepresented in the hydrolysate duodenal digests, but several related peptides found in the casein samples showed a different cleavage pattern with proline–proline at the *N*-terminal end (Figure 5C). The peptide known as neocasomorphin ^{114}YPVEPF^{119} was found in duodenal digests from both casein and hydrolysate, although several peptides lacking the *N*-terminal tyrosine were especially abundant in the hydrolysate samples (Figure 5D). Because the presence of tyrosine at the *N*-terminal end is a structural requirement for opioid receptor agonists, the loss of this amino acid in the hydrolysate digests suggested a decrease in their opioid effect. Other regions where differences between substrates were notable corresponded to α_{s1}-casein 154–173 (Figure 5E) and the α_{s1}-casein *N*-terminal region (Figure 5F).

The analysis of the cleavage positions for both samples was carried out with ggseqlogo to visualize the most likely amino acid found at each of the three *N*- and *C*-terminal positions for both substrates (Figure S4, Supplementary Material). While for casein digests, the most probably *N*-terminal residue is a tyrosine, for hydrolysate digests, it is proline followed by tyrosine. Similarly, the most abundant *C*-terminal tripeptide is PPL and PQQ for casein and hydrolysate, respectively.

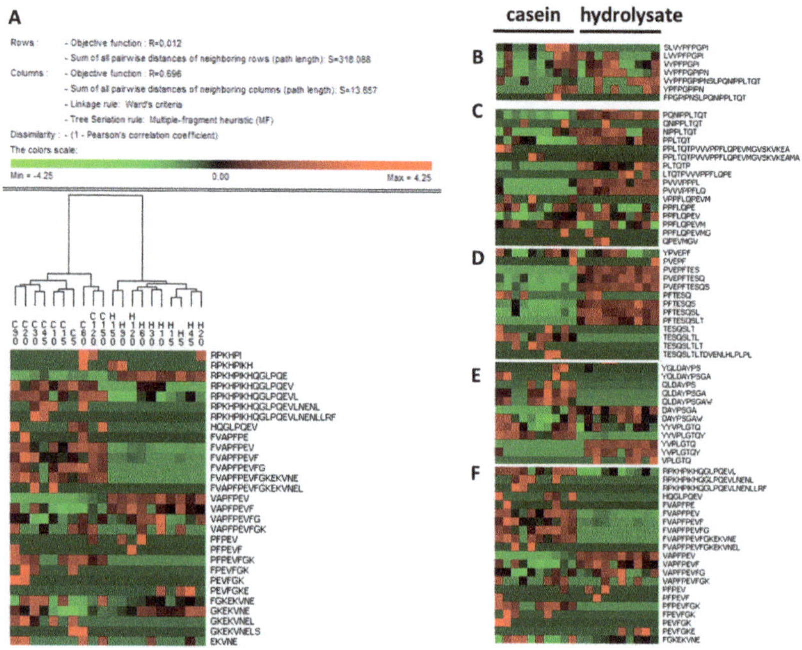

Figure 5. (**A**) Heat map representation (fragment) of the hierarchical clustering analysis of peptides identified in pig duodenal digests (n = 6) at all collected times using the Permut software. Columns correspond to collected times for substrates casein (C) and hydrolysate (H). Selection of areas from the complete representation in panels (**B**–**F**). The complete high-resolution image can be found in .png format in Supplementary Material.

Plasma amino acids were also quantified over time of digestion. In addition to a baseline sample, blood was taken 5, 20, 45, 60, 120, 180, 240 and 360 min after test meal intake. As expected, a greater increase in total amino acids in plasma was observed with the hydrolysate from 45 to 240 min (Figure S5, Supplementary Material), but the values converged at 360 min. Previous reports in humans have described a faster rate of absorption with hydrolyzed casein than with intact casein, although similar whole-body postprandial retention of dietary nitrogen was observed [33]. Moreover, hydrolyzed casein did not enhance ileal endogenous protein losses in humans when compared with the parent protein [34]. When considering the dairy matrix of iso-protein products in the postprandial blood amino acid concentrations in human volunteers, the effect of the protein structure was evidenced as extended response curves with a micellar casein isolate as opposed to a whey protein isolate and yoghurt [35].

Figure 6 shows the determinations of selected amino where significant differences were shown at short times in the case of methionine, lysine or phenylalanine, peaking at 45 min. Other amino acids and metabolites are shown in the Supplementary Material Figure S6. Differences in the 2–3 h post-ingestion period were observed for valine, isoleucine, proline, and two metabolites: α-aminoadipic and α-aminobutyric acids. These values are in accordance with previous results in humans [33] where indispensable amino acids were significantly higher at short times after hydrolyzed casein intake compared to intact casein.

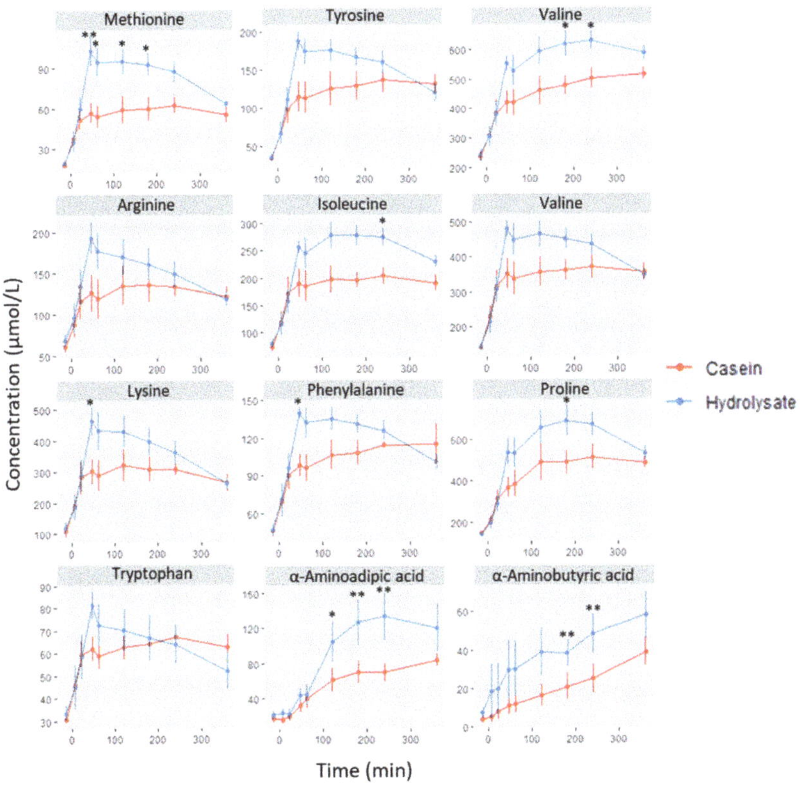

Figure 6. Concentration of amino acids and metabolites in plasma of pigs (n = 6) after intake of casein or hydrolysate over time at basal (−15 min), 5, 20, 45, 60, 120, 180, 240, and 360 min. Differences between substrates at each time point and amino acid were found through a two-way ANOVA, followed by the Bonferroni test: ** $p < 0.01$; * $p < 0.05$.

4. Conclusions

Our results confirm the emptying of intact casein over a more extended period, compared to casein hydrolysate. Discriminant analysis showed a markedly different peptide profile of duodenal digests for all casein fractions between both substrates. In the opioid peptide β-casomorphin-7, precursors with differing terminal ends were observed, with a prevalence of the VYPF- and YPF-ends in the hydrolysate, whereas some longer precursors were more abundant in the casein digests. For other opioid peptides, such as neocasomorphin ^{114}YPVEPF^{119}, an inactive form lacking the *N*-terminal tyrosine, prevailed in the hydrolysate samples. The peptide profile at different time points within the same substrate was similar, suggesting that the degree of protein degradation of the material reaching the duodenum is similar. Finally, the amino acid concentrations found in plasma confirmed a more rapid absorption in hydrolysate-fed animals, with several branched-chain and other hydrophobic amino acids peaking at shorter times than casein. These results can be applied to dairy ingredients formulated as hydrolysates or partly hydrolyzed products, such as infant formula or products used for immune, metabolic or hormonal effects.

Supplementary Materials: The following supporting information can be downloaded at: https://www.mdpi.com/article/10.3390/nu15051065/s1, Figure S1: Electrophoretic profile of pig duodenal effluents at different times of digestion after intake of casein and hydrolysate compared with the given substrate; Figure S2: Identified peptides in pig duodenal digests after intake of casein or hydrolysate over collection time represented by area graphs; Figure S3: Alignment of peptides identified in pig duodenal effluents collected at different times of digestion after intake of casein or hydrolysate and represented as heatmaps of β-casein and α_{S1}-casein; Figure S4: SeqLogo analysis of terminal amino acid residues corresponding to the peptides identified in pig duodenal effluents at all times of digestion after intake of casein or hydrolysate; Figure S5: Total amino acids and metabolites determined in plasma after intake of casein or hydrolysate in the pig model; Figure S6: Concentration of amino acids in plasma of pigs (n = 6) at baseline (−15 min) and 5, 20, 45, 60, 120, 180, 240 and 360 min after intake of casein or hydrolysate. Table S1: Pig endogenous proteins identified in the analyzed duodenal effluents. Full PermutMatrix results are shown as an attached figure in .png format.

Author Contributions: Conceptualization, I.R. and B.M.; methodology, Y.L.G. and P.J.-B.; software, P.J.-B.; formal analysis, P.J.-B., L.S.-R. and D.M.-M.; investigation, P.J.-B., L.S.-R., D.D. and D.M.-M.; resources, D.D., Y.L.G. and I.R.; data curation, P.J.-B.; writing—original draft preparation, P.J.-B. and I.R.; writing—review and editing, P.J.-B., B.M. and I.R.; visualization, P.J.-B.; supervision, B.M. and I.R.; project administration, B.M. and I.R.; funding acquisition, I.R. All authors have read and agreed to the published version of the manuscript.

Funding: This work has received financial support from grants PID2019-107663RB-I00 and PDC2022-133489-I00 funded by MCIN/AEI/10.13039/501100011033.

Institutional Review Board Statement: The local Committee for Ethics in Animal Experimentation approved the study (CE15-10, Aubière, France).

Informed Consent Statement: Not applicable.

Data Availability Statement: Not applicable.

Conflicts of Interest: The authors declare no conflict of interest.

References

1. Boirie, Y.; Dangin, M.; Gachon, P.; Vasson, M.-P.; Maubois, J.-L.; Beaufrère, B. Slow and Fast Dietary Proteins Differently Modulate Postprandial Protein Accretion. *Proc. Natl. Acad. Sci. USA* **1997**, *94*, 14930–14935. [CrossRef] [PubMed]
2. Barbé, F.; Ménard, O.; Gouar, Y.L.; Buffière, C.; Famelart, M.-H.; Laroche, B.; Feunteun, S.L.; Rémond, D.; Dupont, D. Acid and Rennet Gels Exhibit Strong Differences in the Kinetics of Milk Protein Digestion and Amino Acid Bioavailability. *Food Chem.* **2014**, *143*, 1–8. [CrossRef] [PubMed]
3. Vandenplas, Y.; Alarcon, P.; Fleischer, D.; Hernell, O.; Kolacek, S.; Laignelet, H.; Lönnerdal, B.; Raman, R.; Rigo, J.; Salvatore, S.; et al. Should Partial Hydrolysates Be Used as Starter Infant Formula? A Working Group Consensus. *J. Pediatr. Gastroenterol. Nutr.* **2016**, *62*, 22–35. [CrossRef] [PubMed]
4. Jonker, J.T.; Wijngaarden, M.A.; Kloek, J.; Groeneveld, Y.; Gerhardt, C.; Brand, R.; Kies, A.K.; Romijn, J.A.; Smit, J.W.A. Effects of Low Doses of Casein Hydrolysate on Post-Challenge Glucose and Insulin Levels. *Eur. J. Intern. Med.* **2011**, *22*, 245–248. [CrossRef]
5. Drummond, E.; Flynn, S.; Whelan, H.; Nongonierma, A.B.; Holton, T.A.; Robinson, A.; Egan, T.; Cagney, G.; Shields, D.C.; Gibney, E.R.; et al. Casein Hydrolysate with Glycemic Control Properties: Evidence from Cells, Animal Models, and Humans. *J. Agric. Food Chem.* **2018**, *66*, 4352–4363. [CrossRef]
6. Shen, J.; Mu, C.; Wang, H.; Huang, Z.; Yu, K.; Zoetendal, E.G.; Zhu, W. Stimulation of Gastric Transit Function Driven by Hydrolyzed Casein Increases Small Intestinal Carbohydrate Availability and Its Microbial Metabolism. *Mol. Nutr. Food Res.* **2020**, *64*, 2000250. [CrossRef]
7. Dalziel, J.; Young, W.; McKenzie, C.; Haggarty, N.; Roy, N. Gastric Emptying and Gastrointestinal Transit Compared among Native and Hydrolyzed Whey and Casein Milk Proteins in an Aged Rat Model. *Nutrients* **2017**, *9*, 1351. [CrossRef]
8. Mihatsch, W.; Högel, J.; Pohlandt, F. Hydrolysed Protein Accelerates the Gastrointestinal Transport of Formula in Preterm Infants. *Acta Paediatr.* **2001**, *90*, 196–198. [CrossRef]
9. Mihatsch, W.A.; Franz, A.R.; Kuhnt, B.; Högel, J.; Pohlandt, F. Hydrolysis of Casein Accelerates Gastrointestinal Transit via Reduction of Opioid Receptor Agonists Released from Casein in Rats. *Neonatology* **2005**, *87*, 160–163. [CrossRef]
10. Picariello, G.; Mamone, G.; Nitride, C.; Addeo, F.; Ferranti, P. Protein Digestomics: Integrated Platforms to Study Food-Protein Digestion and Derived Functional and Active Peptides. *TrAC Trends Anal. Chem.* **2013**, *52*, 120–134. [CrossRef]
11. Caira, S.; Picariello, G.; Renzone, G.; Arena, S.; Troise, A.D.; De Pascale, S.; Ciaravolo, V.; Pinto, G.; Addeo, F.; Scaloni, A. Recent Developments in Peptidomics for the Quali-Quantitative Analysis of Food-Derived Peptides in Human Body Fluids and Tissues. *Trends Food Sci. Technol.* **2022**, *126*, 41–60. [CrossRef]
12. Basilicata, M.G.; Pepe, G.; Sommella, E.; Ostacolo, C.; Manfra, M.; Sosto, G.; Pagano, G.; Novellino, E.; Campiglia, P. Peptidome Profiles and Bioactivity Elucidation of Buffalo-Milk Dairy Products after Gastrointestinal Digestion. *Food Res. Int.* **2018**, *105*, 1003–1010. [CrossRef] [PubMed]

13. Grosvenor, A.J.; Haigh, B.J.; Dyer, J.M. Digestion Proteomics: Tracking Lactoferrin Truncation and Peptide Release during Simulated Gastric Digestion. *Food Funct.* **2014**, *5*, 2699–2705. [CrossRef] [PubMed]
14. Sánchez-Rivera, L.; Ménard, O.; Recio, I.; Dupont, D. Peptide Mapping during Dynamic Gastric Digestion of Heated and Unheated Skimmed Milk Powder. *Food Res. Int.* **2015**, *77*, 132–139. [CrossRef]
15. Cattaneo, S.; Stuknytė, M.; Masotti, F.; De Noni, I. Protein Breakdown and Release of β-Casomorphins during in Vitro Gastro-Intestinal Digestion of Sterilised Model Systems of Liquid Infant Formula. *Food Chem.* **2017**, *217*, 476–482. [CrossRef] [PubMed]
16. Boutrou, R.; Gaudichon, C.; Dupont, D.; Jardin, J.; Airinei, G.; Marsset-Baglieri, A.; Benamouzig, R.; Tomé, D.; Leonil, J. Sequential Release of Milk Protein Derived Bioactive Peptides in the Jejunum in Healthy Humans. *Am. J. Clin. Nutr.* **2013**, *97*, 1314–1323. [CrossRef]
17. Sanchón, J.; Fernández-Tomé, S.; Miralles, B.; Hernández-Ledesma, B.; Tomé, D.; Gaudichon, C.; Recio, I. Protein Degradation and Peptide Release from Milk Proteins in Human Jejunum. Comparison with in Vitro Gastrointestinal Simulation. *Food Chem.* **2018**, *239*, 486–494. [CrossRef]
18. Egger, L.; Schlegel, P.; Baumann, C.; Stoffers, H.; Guggisberg, D.; Brügger, C.; Dürr, D.; Stoll, P.; Vergères, G.; Portmann, R. Physiological Comparability of the Harmonized INFOGEST in Vitro Digestion Method to in Vivo Pig Digestion. *Food Res. Int.* **2017**, *102*, 567–574. [CrossRef]
19. Contreras, M.d.M.; Sevilla, M.A.; Monroy-Ruiz, J.; Amigo, L.; Gómez-Sala, B.; Molina, E.; Ramos, M.; Recio, I. Food-Grade Production of an Antihypertensive Casein Hydrolysate and Resistance of Active Peptides to Drying and Storage. *Int. Dairy J.* **2011**, *21*, 470–476. [CrossRef]
20. Miralles, B.; Sanchón, J.; Sánchez-Rivera, L.; Martínez-Maqueda, D.; Le Gouar, Y.; Dupont, D.; Amigo, L.; Recio, I. Digestion of Micellar Casein in Duodenum Cannulated Pigs. Correlation between in Vitro Simulated Gastric Digestion and in Vivo Data. *Food Chem.* **2021**, *343*, 128424. [CrossRef]
21. Wickham, H. *Ggplot2 Elegant Graphics for Data Analysis*; Use R! Springer: New York, NY, USA, 2016; ISBN 978-3-319-24277-4.
22. Wagih, O. Ggseqlogo: A Versatile R Package for Drawing Sequence Logos. *Bioinformatics* **2017**, *33*, 3645–3647. [CrossRef]
23. Caraux, G.; Pinloche, S. PermutMatrix: A Graphical Environment to Arrange Gene Expression Profiles in Optimal Linear Order. *Bioinformatics* **2005**, *21*, 1280–1281. [CrossRef]
24. Mahe, S.; Roos, N.; Benamouzig, R.; Rautureau, J.; Tome, D. Gastrojejunal Kinetics and the Digestion of [15N] F3- Lactoglobulin and Casein in Humans: The Influence of the Nature and Quantity of The. *Am. J. Clin. Nutr.* **1996**, *63*, 546–552. [CrossRef] [PubMed]
25. Barbé, F.; Le Feunteun, S.; Rémond, D.; Ménard, O.; Jardin, J.; Henry, G.; Laroche, B.; Dupont, D. Tracking the in Vivo Release of Bioactive Peptides in the Gut during Digestion: Mass Spectrometry Peptidomic Characterization of Effluents Collected in the Gut of Dairy Matrix Fed Mini-Pigs. *Food Res. Int.* **2014**, *63 Pt B*, 147–156. [CrossRef]
26. Picariello, G.; Miralles, B.; Mamone, G.; Sánchez-Rivera, L.; Recio, I.; Addeo, F.; Ferranti, P. Role of Intestinal Brush Border Peptidases in the Simulated Digestion of Milk Proteins. *Mol. Nutr. Food Res.* **2015**, *59*, 948–956. [CrossRef] [PubMed]
27. Asledottir, T.; Le, T.T.; Petrat-Melin, B.; Devold, T.G.; Larsen, L.B.; Vegarud, G.E. Identification of Bioactive Peptides and Quantification of β-Casomorphin-7 from Bovine β-Casein A1, A2 and I after Ex Vivo Gastrointestinal Digestion. *Int. Dairy J.* **2017**, *71*, 98–106. [CrossRef]
28. Brantl, V.; Teschemacher, H.; Henschen, A.; Lottspeich, F. Novel Opioid Peptides Derived from Casein (Beta-Casomorphins). I. Isolation from Bovine Casein Peptone. *Hoppe Seylers Z Physiol. Chem.* **1979**, *360*, 1211–1216. [CrossRef] [PubMed]
29. Jinsmaa, Y.; Yoshikawa, M. Enzymatic Release of Neocasomorphin and [Beta]-Casomorphin from Bovine [Beta]-Casein. *Peptides* **1999**, *20*, 957–962. [CrossRef] [PubMed]
30. Loukas, S.; Varoucha, D.; Zioudrou, C.; Streaty, R.A.; Klee, W.A. Opioid Activities and Structures of.Alpha.-Casein-Derived Exorphins. *Biochemistry* **1983**, *22*, 4567–4573. [CrossRef]
31. Fernández-Tomé, S.; Martínez-Maqueda, D.; Girón, R.; Goicoechea, C.; Miralles, B.; Recio, I. Novel Peptides Derived from As1-Casein with Opioid Activity and Mucin Stimulatory Effect on HT29-MTX Cells. *J. Funct. Foods* **2016**, *25*, 466–476. [CrossRef]
32. Nakamura, Y.; Yamamoto, N.; Sakai, K.; Takano, T. Antihypertensive Effect of Sour Milk and Peptides Isolated from It That Are Inhibitors to Angiotensin I-Converting Enzyme. *J. Dairy Sci.* **1995**, *78*, 1253–1257. [CrossRef] [PubMed]
33. Deglaire, A.; Fromentin, C.; Fouillet, H.; Airinei, G.; Gaudichon, C.; Boutry, C.; Benamouzig, R.; Moughan, P.J.; Tomé, D.; Bos, C. Hydrolyzed Dietary Casein as Compared with the Intact Protein Reduces Postprandial Peripheral, but Not Whole-Body, Uptake of Nitrogen in Humans. *Am. J. Clin. Nutr.* **2009**, *90*, 1011–1022. [CrossRef] [PubMed]
34. Deglaire, A.; Moughan, P.J.; Airinei, G.; Benamouzig, R.; Tomé, D. Intact and Hydrolyzed Casein Lead to Similar Ileal Endogenous Protein and Amino Acid Flows in Adult Humans. *Am. J. Clin. Nutr.* **2020**, *111*, 90–97. [CrossRef] [PubMed]
35. Horstman, A.M.H.; Ganzevles, R.A.; Kudla, U.; Kardinaal, A.F.M.; van den Borne, J.J.G.C.; Huppertz, T. Postprandial Blood Amino Acid Concentrations in Older Adults after Consumption of Dairy Products: The Role of the Dairy Matrix. *Int. Dairy J.* **2021**, *113*, 104890. [CrossRef]

nutrients

MDPI

Article

Impact of Grape Seed Powder and Black Tea Brew on Lipid Digestion—An In Vitro Co-Digestion Study with Real Foods

Judit Tormási and László Abrankó *

Department of Food Chemistry and Analytical Chemistry, Institute of Food Science and Technology, Hungarian University of Agriculture and Life Sciences (MATE), Villányi Street 29-43, 1118 Budapest, Hungary
* Correspondence: abranko.laszlo.peter@uni-mate.hu

Abstract: Effects of two foods with bioactive constituents (black tea brew, BTB and grape seed powder, GSP) on lipid digestibility was studied. Lipolysis inhibitory effect of these foods was examined using two test foods (cream and baked beef) with highly different fatty acid (FA) composition. Digestion simulations were performed either using both gastric and pancreatic lipase, or only with pancreatic lipase according to the Infogest protocol. Lipid digestibility was assessed based on the bioaccessible FAs. Results showed the triacylglycerols containing short- and medium-chain FAs (SCFA and MCFA) are non-preferred substrates for pancreatic lipase; however, this is not characteristic for GL. Our findings suggest that both GSP and BTB primarily affect the lipolysis of SCFAs and MCFAs, because the dispreference of pancreatic lipase towards these substrates was further enhanced as a result of co-digestion. Interestingly, GSP and BTB similarly resulted in significant decrease in lipolysis for cream (containing milk fat having a diverse FA profile), whereas they were ineffective in influencing the digestion of beef fat, having simpler FA profile. It highlights that the characteristics of the dietary fat source of a meal can be a key determinant on the observed extent of lipolysis when co-digested with foods with bioactive constituents.

Keywords: lipid digestibility; milk fat; beef fat; gastric lipase; grape seed powder; black tea; bioaccessibility

1. Introduction

Obesity and associated diseases such as metabolic syndrome, cardiovascular disease, type 2 diabetes, are unarguably among the most important health concerns [1–4]. The imbalance between the calories consumed with food and the energy expenditure of the body plays an important role in the development of obesity [5]. The contributions of both total dietary fat intake and percentage of energy intake from fat to obesity were shown by many studies [6,7]. However, others also indicated that this observation cannot always be confirmed at the population level [8] or divergencies were found for some sociodemographic groups [9]. These findings may support the rather fatty acid-dependent than generic associations between dietary fat intake and body weight, which is also shown in a number of studies [10,11].

Dietary fats—mainly comprised of triacylglycerols (TAGs)—are at least partly hydrolyze to free fatty acids (FFAs) via enzymatic hydrolysis during human digestion. At the end of lipid digestion, FFAs and monoacylglycerols (MAGs) are counted as bioaccessible molecules from which after trough passive or active transport part are transferred into the bloodstream, providing bioavailable molecules for further mechanisms [12–14]. The two main enzymes guiding the breakdown of food lipids are gastric lipase (GL) and pancreatic lipase (PL). The first step takes place in the stomach and is performed by GL; however, the majority of the lipid digestion by PL in the duodenum [15]. In general, GL contributes to approximately 5–40% of overall lipid digestion [16]. In addition, GL also serves as a catalyst for further lipid digestion, i.e., pre-digestion and emulsification enhances the efficacy of pancreatic lipase digestion [17]. Following the gastric pre-digestion, the food bolus enters the duodenum where PL continues the lipolysis. In total, more than 50% of the overall lipolysis is performed by PL in the small intestine. TAG hydrolysis—both performed by GL or PL—results in release of FFAs and partly hydrolyzed products such as MAGs, and diacylglycerols (DAG). It is important to note that unhydrolyzed TAGs are also present during the digestion process.

Among the many approaches applied in clinical obesity therapy, drug treatments using lipase inhibitors became an important means of clinical obesity treatment [18]. By restricting the liberation of fatty acids (FA), the bioaccessibility of dietary lipids can be modulated [19]. In Europe and in the US, currently, the only authorized medicine for lipase inhibition used in clinical treatment is Orlistat (lipstatin, Xenical®).

Alternatively, natural lipase inhibitors present in our food commodities are consumed over longer periods of time as a part of our everyday diet, and they can also play an important role in reduction in dietary lipid bioaccessibility and, consequently, preventing obesity [19–21]. Promising examples of foods with lipase inhibitory effect were reported spices, e.g., rosemary [22] and sage [23], industrial waste products such as pomegranate leaf [24] or grape seed [25,26], and commonly consumed beverages such as tea cultivars [27,28].

Wine production is one of the most important agricultural activities in the world and causes the generation of a large number of by-products, including grape skins, -seeds, and -stems [29]. Wine by-products have high potential as food ingredients, since they facilitate increased sustainability increase in the wine industry by reusing a product that is usually considered waste [30]. Therefore, further utilization of by-products can be of great importance, and a great deal of research effort is being devoted to testing the putative beneficial effects of grape parts [31]. Both grape seed [25] and grape skin [32] extract showed potential pancreatic lipase inhibitory effect according to in vitro enzyme assays using model substrates. Lipolysis inhibitory effect is often attributed to a variety of polyphenolic compounds, e.g., proanthocyanidins or condensed tannins, found in solid parts of grape (seeds and skins) [26,33–36]. Lipid profile of grapeseed was also studied and it was found that most abundant fatty acids are linoleic acid (C18:2 ω-6), oleic acid (C18:1), palmitic acid (C16:0), and stearic acid (C18:0), which represent about 98% of the total fatty acids [37].

Tea (*Camellia sinensis*) is one of the most frequently consumed beverages [38]. The health benefits of several types of tea (white [39], oolong [27], Pu-erh [28], green [38] and black [40,41]) were studied, which all are associated with potent lipase inhibitory properties. and these effects are closely related to the structure and composition of polyphenols. Catechins, which, during the fermentation process, are converted to theaflavins and large polymeric compounds, such as thearubigins, are the characteristic polyphenols in black tea brew [42].

It should be noted that the mechanism of action of different agents resulting in a reduction in lipid bioaccessibility can be highly different. Some of them—such as Orlistat—can bind directly to the active site of PL, whereas natural compounds are more likely to act indirectly, e.g., by interacting the oil–water interface [43]. Moreover, the role of polyphenols in terms of synergisms with other non-polyphenol constituents and their effect on the interface microenvironment is still inconclusive. It suggests that despite the available evidence on the role of polyphenols in lipase inhibition, it may be misleading at this stage of our knowledge to attribute the effect of polyphenol-rich foods such as grape seed powder or tea brew on lipolysis exclusively to polyphenols. Therefore, it would be desirable to study the putative effects of such bioactive components of foods such as polyphenols in their natural matrix or as part of a meal, since beyond direct lipase inhibition, food composition and co-digestion of other foods are also considered as key determinants of the extent of lipolysis.

The effect of food matrix, lipid content and composition on bioaccessibility and bioavailability of bioactive substances using the Infogest in vitro digestion simulation were studied and reviewed [44–46]. These studies highlight the importance of the presence of the lipid molecules, the sufficient emulsification and lipid hydrolysis during digestion on the release of bioactive molecules. In addition, it was reported that oils with different composition might show different overall bioaccessibility in presence of the same bioactive molecule, such as β-carotene or eugenol [47,48]. However, these results are based on the use of model matrices and there is still a lack of studies providing data on lipid digestibility from real foods and on how it is affected by bioactive substances.

Although there are some exceptions where more complex digestion simulation is used [49], most studies still focus on use of enzyme assays with model substrates and the diversity in the lipid composition of consumed food commodities is often neglected. Based on the encouraging results of the enzyme assays on the influence of grape seed powder (GSP) and black tea brew (BTB) on lipid digestion, the primary aim of this work was to evaluate the impact of these two selected foods on lipid digestion when co-digested with real foods with diverse lipid profile, under conditions that closely simulates human digestion [50].

2. Materials and Methods

2.1. Materials

For digestion protocol: porcine α-amylase (EC 3.2.1.1), porcine pepsin (EC 3.4.23.1), porcine pancreatin (8 × USP; P7545) and porcine bile extract (E.C. 3.1.1.3) was purchased from Merck/Sigma-Aldrich, St. Louis, MO, USA. Rabbit gastric lipase (RGE15) was purchased from Lipolytech Ltd., Marseille, France. FAME mixture (CRM47885) and internals standards; glyceryl trinonadecanoate (C19:0 TAG; >99%), methyl nonadecanoate (C19:0 ME, analytical standard), glyceryl triheptadecanoate (C17:0 TAG) and heptadecanoic acid (C17:0 FFA) were also obtained from Merck/Sigma-Aldrich. L-serine standard was purchased from Reanal Ltd., Budapest, Hungary Reagents and solvents were of analytical purity. Chloroform was purchased from Carlo Erba Reagents, Emmendingen, Germany (for analysis, stabilized with ethanol), methanol (for HPLC, LC-MS grade) and hydrochloric acid (37%) from VWR International, Radnor, PA, USA, isooctane from Fisher Scientific, Waltham, MA, USA (>95%). High purity water (>18 MΩcm^{-1}) was prepared by a Millipore Elix Essential 3 UV Water Purification System (Merck-Millipore, Burlington, MA, USA). Orlistat (>98% solid) were purchased from Sigma-Aldrich.

2.2. Sample Preparation

Substrate test foods, ground beef (20% fat content) and cream (30% fat content) were purchased from local supermarkets. Ground beef was baked before digestion experiments for 20 min in a 200 °C oven and it was then homogenized in a household meat grinder three times. Baked beef (BB) samples were stored at −80 °C and thawed before experiments. Cream (C) was always purchased fresh and used right after opening and thorough mixing on room temperature. These substrate test foods were selected as representatives of natural dietary lipids sources to gain information on the effect of fat-free foods; grape seed powder and black tea on lipid digestion.

Grape seed powder (GSP) was provided by Bock Vineyard Ltd. (Villány, Hungary), and the powder was added to substrate test foods (BB and C) in multiple concentrations before digestion experiments, separately. Black tea (Himalayan Spring FF 2022 No. 601) with high tannin content were bought commercially. Tannin content (163 mg/g tea leaf) was measured according to the method provided for determination of tannin content of tea species [51]. The same extract as to determine tannin content was used in digestion experiments, as black tea brew (BTB). BTB was added to digestion simulations test foods (baked beef and cream) in multiple concentrations before the experiments.

Determination of tannin content was as follows. Into 50-mL round bottom flasks, 0.2000 g of black tea leaves were measured and 50 mL distilled water was added. Water was heated on sand for 1 h (with water cooler system attached to prevent evaporation). After cooling, extracts were sieved on paper sieve and collected filtrate was completed to 50 mL. One part of the extract was used to determine tannin content, the rest was stored in −80 °C until use in digestion experiments. For measuring tannin content, extract was heated until boiling and 30 mL 5 $w/v\%$ copper–acetate solution was added to precipitate tannin content. Extract was sieved and pellets were washed until no residual copper was left. Pellets were cremated, rewetted with nitric acid and re-cremated. Tannin content was determined by weight.

2.3. *Determination of Fat Content of Test Foods*

For evaluating total fat content of baked beef, the standard method available for meat and meat products was used [52] based on Soxhlet extraction with petroleum ether, and gravimetric determination. Total fat content of cream was determined according to ISO 2450:2009 [53] using cold, organic extraction with the mixture of diethyl- and petroleum ether, and gravimetric determination. Total fat contents of test foods were given in g fat/100 g product. Fatty acid composition was determined from the extracted fats with the GC-FID method described in Tormási and Abrankó [54].

2.4. *In Vitro Digestion Simulation*

Digestion simulations were made according to Infogest internationally acknowledged consensus in vitro digestion simulation protocol [50]. The original 2014 version of this protocol did not include gastric lipase, whereas the most recent version suggests to also use gastric lipase (as a part of rabbit gastric extract) when lipid digestion is to be studied. To better understand the synergistic effect of GL in the overall lipid digestions by performing pre-digestion, both versions of the protocol were used. In the version hereafter called GL + PL, lipolytic enzymes gastric lipase (as a part of rabbit gastric extract (RGE) also containing pepsin) and pancreatic lipase (as a part of pancreatin enzyme complex) were used. In the second version, hereafter called PL, only pancreatic lipase (as a part of pancreatin enzyme complex) was used.

Simulated stock electrolyte solutions (simulated salivary fluid; SSF, simulated gastric fluid; SGF, simulated intestinal fluid; SIF) were made beforehand, and pH was adjusted to 7 (oral), 3 (gastric) and 7 (small intestinal), respectively. Enzyme activities were measured via methods described in Brodkorb et al., and enzyme stock solutions were made right before use. Required volume of 6 M HCl and 1 M NaOH to keep pH was determined before each digestion experiment with a "pH test" using the same amounts of samples and solvents only without enzymes.

For oral phase, 3.5 mL of SSF (tempered to 37 °C), 25 μL of 0.3 M $CaCl_2$, 0.5 mL of amylase solution (1500 U/mL in SSF) and 0.975 mL of distilled water was added. Homogenized samples were incubated in an overhead shaker (Heidolph Reax 2, Heidolph Instruments, Schwabach, Germany) fitted inside a preheated drying cabinet (Memmert UNE300, Memmert GmbH, Schwabach, Germany) for 2 min at 37 °C. In the gastric phase, 6.4 mL of SGF (tempered to 37 °C), 5 μL of 0.3 M $CaCl_2$, required volume of 6 M HCl, 1.6 mL of pepsin solution (25,000 U/mL in SGF) and 1.945 mL of water was added, and the mixture was incubated in the overhead shaker at 37 °C for 2 h. In the case of the GL + PL digestion, 1.6 mL of rabbit gastric extract (RGE, 750 U/mL in SGF) was added in the gastric phase and no pepsin was added. For the small intestine phase, 8.5 mL of SIF (tempered to 37 °C), 40 μL of 0.3 M $CaCl_2$, 2.5 mL of bile extract solution (160 mM in SIF, tempered to 37 °C), 5 mL of pancreatin solution (800 U/mL in SIF), required volume of 1 M of NaOH and 3.86 mL of water was added, and mixture was incubated in the overhead shaker at 37 °C another 2 h. After small intestinal digestion phase was completed, weight of digests was measured.

In order to accurately characterize the effect of bioactive containing foods on lipid digestibility, fat content of the digests' were set to 150 mg; therefore, sample weight varied according to fat content of the test foods. Control digestions (only substrate test foods) were carried out with 0.5 g cream and 0.9 g baked beef based on lipid content. Samples were measured into 50 mL centrifuge tubes and supplemented to 5 g with distilled water. In addition of lipid digestibility test of the two test foods, experiments were carried out by co-digestion of the test foods with two fat-free foods containing bioactive constituents

(GSP, BTB). To determine the appropriate concentration, dose dependency tests of GSP and BTB were carried out with cream test food. To cream (0.5 g), BTB was added at 1:1, 1:2 and 1:3 (w) ratio or GSP was added at 5, 10 and 15 *w*/*w*%. Portions of GSP typically consumed are 5–10 g a day, which is in accordance with the used 5–15 *w*/*w*% concentration if consumed along with a tea drink or a meal containing 50–100 g meat. Portions of test foods with appropriate amounts of either GSP or BTB were homogenized and made up to 5 g in 50-mL test tubes before digestion simulation. Effect of GSP and BTB on BB were only tested at the most efficient level of addition, i.e., 5 *w*/*w*% for GSP and 1:2 ratio for BTB. All digestion experiments were conducted in triplicates. Blank digestions were also made for each triplicate using 5 g of distilled water as sample. Positive control for lipase inhibition studies were made with the addition of Orlistat (a lipase inhibitory drug), where 40 µL of 0.5 M Orlistat solution in DMSO was added to fully inhibit lipases (Results presented elsewhere [54]).

2.5. Assessment of Fatty Acid Specific Lipolysis—Bioaccessible Fatty Acid Content

Bioaccessible FA content and composition were measured according to Tormási and Abrankó, 2021 [54]. From digests, 5 mL samples were taken and all lipid compounds were extracted with the Bligh and Dyer method (1959) [55] after the addition of 250 µL of C19:0 TAG internal standard (ISTD) solution (1 mg/mL in $CHCl_3$). After phase separation, from the lower phase (∑12.5 mL chloroform, containing fat components) 2 × 5 mL was pipetted into two round-bottom flasks and solvent was evaporated using a rotary evaporator. One aliquot was methylated according to the ISO 12966-2:2017 standard's 'general method' to result the total fatty acid content (TFA method) [56]. The other 5-mL aliquot was trans methylated with the alkaline 'fast method' of the same ISO standard to obtain the esterified fatty acid content (EFA method). FA composition was determined with GC-FID (see chapter "Determination of fatty acid composition with GC-FID"). Total fat content (TFA) and esterified fat content (EFA) were used to determine amount of free fatty acids (FFA, Equation (1)), bioaccessibility or release ratio of each fatty acids (RR, Equation (2)) and bioaccessible fatty acid content in % (Equation (3)).

$$\text{FFAi} = \text{TFAi} - \text{EFAi} \quad (1)$$

$$\text{Bioaccessibility of FAi/Release Ratio} = \text{FFAi/TFAi} \quad (2)$$

$$\text{Bioaccessible FA content [\%]} = (\textstyle\sum \text{FFAi}/\sum \text{TFAi}) \times 100 \quad (3)$$

TFA_i: total amount of individual FAs; EFA_i: amount of the same FA in esterified form, FFA_i: amount of the same free FA, ∑FFAi: sum of the amount of each individual FFAs and ∑TFAi: the sum of the amount of each individual FA all in the small intestinal digesta.

2.6. Statistics

Pairwise comparison was carried out in Microsoft Excel using Student's *t* test. Significance was recognized at $p < 0.05$. Comparison of multiple samples was performed using ANOVA test in Microsoft Excel (LTSC MSO (16.0.14332.20279) 32-bit); however, where significant difference was found ($p < 0.05$), Tukey's post hoc test was performed using IBM SPSS Statistics 25. Equality of variances were tested with Levene's test ($p > 0.05$).

3. Results and Discussion

3.1. Substrate Test Foods with Different Fatty Acid and TAG Profile

The impact of black tea brew (BTB) and grape seed powder (GSP) on lipid digestion was studied using two substrate test foods: cream (C) and baked beef (BB). The lipid content of substrate test foods was considered as the substrate for lipolysis. Primarily, we wanted to choose real food samples as substrate test foods and, moreover, ones with highly different FA composition.

Milk fat (lipid source of cream) contains a diverse variety of FAs containing short chain fatty acids (SCFA), medium chain fatty acids (MCFA) and long chain fatty acids (LCFA).

According to FA profiling carried out in this study, in the analyzed cream sample, the most abundant FAs were palmitic acid (C16:0: 33.3 $\pm$ 0.3%), oleic acid (C18:1n-9c: 24.5 $\pm$ 0.3%), stearic acid (C18:0: 11.9 $\pm$ 0.2%) and myristic acid (C14:0: 11.2 $\pm$ 0.1%). It also contained several FAs between 1 and 5 w/w% such as linoleic acid (C18:2n-6c: 4.1 $\pm$ 0.1%), capric acid (C10:0: 3.4 $\pm$ 0.04%), lauric acid (C12:0: 2.8 $\pm$ 0.2%), caproic acid (C6:0: 2.6 $\pm$ 0.03%), elaidic acid (C18:1n-9t: 2.4 $\pm$ 0.3%), palmitoleic acid (C16:1n-7c: 2.2 $\pm$ 0.03%), caprylic acid (C8:0: 1.5 $\pm$ 0.04%) and pentadecanoic acid (C15:0: 1.3 $\pm$ 0.02%) (Figure 1). The overwhelming majority (98.2 $\pm$ 0.1%) of the lipid content of the tested cream sample was made of twelve different FAs. The high degree of FA diversity measured in cream (milk fat) necessarily reflects in the types of TAGs as well. A mixture of TAGs with heterogeneous composition is characteristic of milk fat [57]. Short (C4-C8) chain FAs were largely located at the *sn*-3 position of TAGs next to MCFAs and LCFAs, whereas palmitic acid (C16:0) and oleic acid (C:18:1) were predominantly found in *sn*-2 position; however, the preferred position of the latter and also that of MCFAs depend on the actual size of the TAG [58]. Therefore, milk fat TAGs are considered to contain the most "asymmetric" TAGs among animal fats [59].

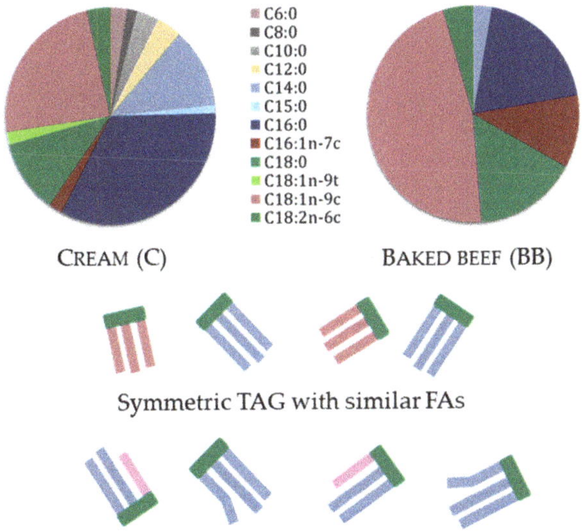

Figure 1. Fat content of test foods cream (C) and baked beef (BB) and illustration of most common TAG structures.

In contrast, the FA profile of beef fat was much simpler than that of milk fat. In the BB test food, only six of the FAs were above 1 w/w% and the majority of them were comprised of only five FAs, namely, oleic acid (C18:1n-9c: 44.9 $\pm$ 0.02%), palmitic acid (C16:0, 26.8 $\pm$ 0.06%), stearic acid (C18:0, 14.9 $\pm$ 0.08%), palmitoleic acid (C16:1n-7c; 5.0 $\pm$ 0.02%), myristic acid (C14:0: 2.6 $\pm$ 0.01%) and linoleic acid (C18:2n-6c, 2.1 $\pm$ 0.02%) (Figure 1). The three most abundant FAs (C18:1n-9c, C16:0 and C18:0) covered the 75.5 $\pm$ 0.21% of all FAs. Consequently, the structure of most beef TAGs was more symmetric than of cream TAGs [60].

3.2. Lipid Digestibility of Substrate Test Foods

Lipid digestibility of studied test foods was assessed based on measuring the extent of lipolysis. This was carried out according to our recently published method [54] based on the monitoring of the bioaccessible FFAs after in vitro digestion simulation.

The applied in vitro digestion simulation protocols allowed us to perform a mechanistic evaluation of the contribution of GL and PL to lipid digestibility of the test foods (cream and baked beef) and, later, of the effect of foods with bioactive constituents (GSP and BTB). Digestion simulations were performed either using both gastric lipase and pancreatic lipase (hereafter GL + PL) or only with pancreatic lipase (hereafter PL). After in vitro digestion simulation with both versions, the FFA content (Equation (1)) and RR (Equation (2)) of each individual FA were determined. Total lipid digestibility was assessed as the sum of individual FFAs (Equation (3)).

3.2.1. Cream: Addition of Gastric Lipase Specifically Ameliorated Release of SFCAs and MCFAs

Total lipid digestibility based on FA release of cream determined with GL + PL method was 77.1 ± 5.0% (RSD: 0.07). The RR of individual FAs slightly varied in the range of 0.70 and 0.82 (Figure 2A). Only one FA (C6:0) deviated from the rest, which showed relatively low digestibility (RR = 0.45) compared to the other FAs. These results show that the FA pattern of the food and the FFA profile of the digesta were similar; thus, no bias in either of the FA occurred during enzymatic hydrolysis of cream, when both lipolytic enzyme types were added.

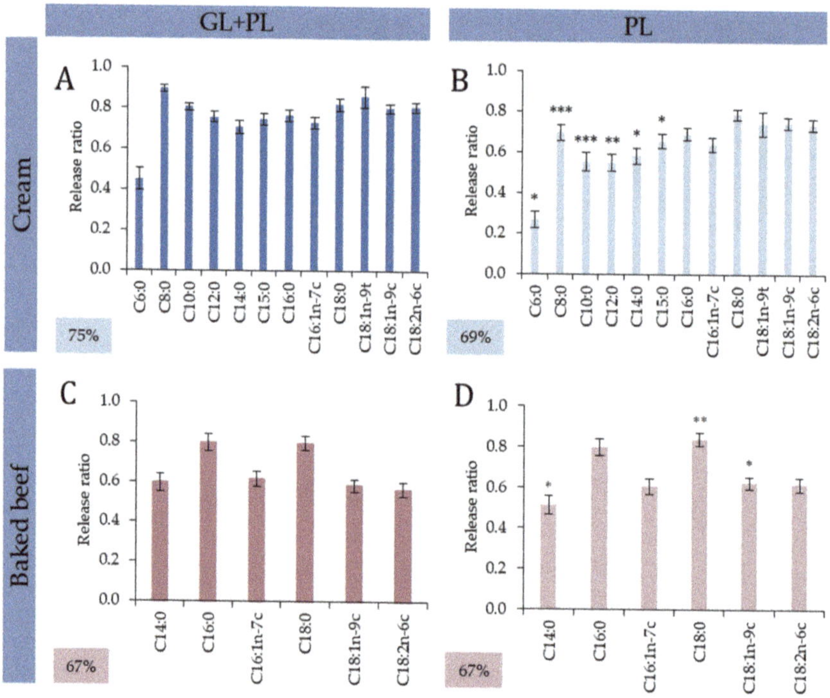

Figure 2. Bioaccessible fatty acid content of test foods (%; Equation (3)) and release ratios (Equation (2)) after Infogest digestion simulations (GL + PL and PL). (**A**): Cream GL + PL, (**B**): Cream PL; (**C**): Baked beef GL + PL, (**D**): Baked beef PL. Asterisk labels (*) show significant difference between release ratio of individual fatty acids after GL + PL and PL digestions (* $p < 0.05$; ** $p < 0.01$; *** $p < 0.001$).

To gain more insight to the different digestion mechanism of lipolysis with and without GL, digestibility of lipids was also measured after in vitro digestion simulation with PL alone. Compared to the results derived from GL + PL digestion, overall digestibility with PL digestion decreased by 11% to 69.2 ± 6.1%. This was in accordance with the literature consensus on the contribution of GL to overall lipid digestion, i.e., pre-digestion by GL adds between 5 and 40% to overall lipid digestion [16].

Nevertheless, the extent of lipolysis was markedly different for shorter chain FAs and longer chain FAs. Short and medium chain FAs showed RR from 0.55 to 0.65 and longer chained FAs showed RR from 0.69 to 0.73. The caproic acid (C6:0) were an outlier during PL digestions as well, RR value was only 0.27 to this FA. The addition of GL mostly affected the release of SCFAs and MCFAs during digestion simulation of cream, namely in the case of C6:0 (+40%), C8:0 (+22%), C10:0 (+31%), C12:0 (+27%), 14:0 (+17%) and C15:0 (+12%) (Figure 2A,B). This can be explained by the difference in the substrate preference of PL and GL. Both GL and PL enzymes have regiospecific preferences for FAs at the outer *sn*-1/3 positions [15,61–63], and it is also generally accepted that GL has a preference for the FAs in the *sn*-3 position in TAGs. However, a lesser preference over the *sn*-1 position was also reported. According to our results, the contribution of GL was most pronounced in the lipolysis of short (C4-C8) and medium (C10-C14) chain FAs. Short chain FAs are mainly located in the *sn*-3 position next to MCFAs (in *sn*-2) in smaller size TAGs [58], most probably forming asymmetric TAGs in milk fat. Our results also lend support to the assumption that that small-sized TAGs containing both SCFA and MCFA are non-preferred substrate types for PL. Consequently, PL alone is rather ineffective in the digestion of TAGs containing short and medium chain FAs and GL is of key importance in the lipolysis of such asymmetric TAGs of milk fat. The findings of Benito-Gallo et al. also suggest that PL showed the same selectivity towards the hydrolysis at positions *sn*-1 and *sn*-3 of the TAG, provided that the FA side chains were identical [64]. This further supports our observations regarding the key role of GL in the digestion of asymmetric TAGs of milk fat. Moreover, due to this shift in RR of these FAs with the addition of GL, the saturated FA level increased by 14% in the bioaccessible fraction of digested cream compared to just using PL.

3.2.2. Baked Beef: No Additional Effect of GL without Diversity of TAGs

In the case of baked beef test food, total FA release and individual RR of FAs were also determined after in vitro digestion simulation. Interestingly, there was no difference between total digestibility measures determined after GL + PL and PL digestions. Overall, FA release was 67.7 $\pm$ 2.5% and 67.2 $\pm$ 1.6%, respectively ($p = 0.742$). In contrast, there were some differences between individual RRs of FAs. With the addition of GL, RR of C14:0 increased from 0.51 to 0.60 ($p = 0.019$) and RR of C18:0 and C18:2n-6c decreased slightly from 0.84 and 0.62 to 0.80 ($p = 0.001$) and 0.56 ($p = 0.036$), respectively (Figure 2C,D). However, the observed changes in the FA composition of the digesta caused by the difference in the RR of these FAs cancelled each other out; thus, no change was detectable in the overall digestibility of baked beef fat. The differences between the digestion simulation with and without GL showed that GL had a significant role in ameliorating and improving the release of certain FAs. Although the digestion simulation did not show variation between overall lipid digestibility of baked beef, the release of one MCFA (C14:0) increased significantly (+14%) and two others decreased (both LCFAs, C18:0 and C18:2n-6c; by 5–9%). It should be noted, however, that in the case of baked beef, FA composition of digesta differed from FA composition of the baked beef meal. It means, the relative amount of C14:0, C16:1n-7c, C18:1n-9c and C18:2n-6c decreased and C16:0 and C18:0 increased in the digesta compared to the test food; thus, SFA content of digesta also increased by 13.7%. Similarly, Hur et al. investigated the lipid digestibility of beef and, in line with our observations, they found that after in vitro digestion, the composition of digesta differed from initial FA composition of beef [65].

The results of our fat digestibility study on cream (milk fat) and baked beef showed the TAGs containing SCFAs and MCFAs were non-preferred substrate types for PL and, consequently, PL alone was rather ineffective in the digestion of such TAGs containing SCFAs and MCFAs. However, this specificity is not characteristic for GL. Thus, the presence of GL resulted in the increased release of SCFAs and MCFAs (C6:0–C15:0), especially if these FAs were part of asymmetric TAGs. This effect was more pronounced in milk fat containing products such as the analyzed cream. The results also showed that when there was no marked diversity in FA composition and, consequently, heterogeneity in the TAG structure is limited, the contribution of GL could be negligible such as in the case of baked beef, where overall fat digestibility remained the same. However, the contribution of GL to fat digestibility might be more complex in other food commodities. It was shown that the ability of GL to cleave the PL-not preferred FAs from *sn*-3 position would have additional benefits. The TAGs that are ineffectively digested by PL—such as presumably asymmetric TAGs containing SCFAs and MCFAs—would appear in pre-digested *sn*-1,2/2,3 DAG forms in the next (ileal) stage of digestion, which PL can now cope with more easily. Moreover, the FFAs produced during gastric digestion, due to their emulsifying ability can further improve PL's access to TAGs, thus improving the efficiency of fat digestion in the duodenum [17]. This synergistic effect of the two enzymes is clearly demonstrated in lipid digestion of substrates with substantial quantity of SCFAs and MCFAs, such as cream. Finally, it should be also emphasized that the observed variation of the lipolytic efficiency of GL and PL towards different TAGs can also be attributed to differences in the interfacial microenvironment of the enzymes. This aspect can be of special importance during real meal digestion processes, when a number of uncontrolled interfacial interactions may occur [66].

3.3. *Effect of GSP and BTB on Lipid Digestibility*

Two different types of fat-free foods with bioactive constituents with proven lipase inhibitory effects, namely grape seed powder (GSP) and black tea brew (BTB), were selected to investigate the characteristics of their effects on the digestion of milk fat (cream) and baked beef fat. First, effective treatment levels of GSP and BTB were determined by performing dose–response experiments on cream as sample test foods. The effective dosage was further tested on BB for evaluation of substrate-specific effects.

3.3.1. Dose and Substrate Dependency

To evaluate the required effective dosage of the selected foods with bioactive constituents, first, co-digestion experiments with cream test food were conducted at three levels. In the case of GSP, which was available as a powder, it was tested in small quantities, namely in 5, 10 and 15 *w*/*w*% relative to the weight of cream. As Figure 3A shows, a significant decrease was observed at the first level (5% GSP) and the further addition of GSP did not result in a decrease in lipid digestibility. The same was noticed in both GL + PL and PL digestion simulations, i.e., the lowest tested level already resulted a significant effect in cream. The observed decrease in lipolysis in PL digestions indicates that the dispreference of PL to digest TAGs containing SCFAs and MCFAs was presumably further enhanced. At the same time, the results of the GL + PL digestions suggest that GSP had an impact on GL; thus, the efficacy of the lipolysis of TAGs containing SCFAs and MCFAs was decreased, i.e., these TAGs continue to reach the duodenum. Their lipolysis will not be completed in the small intestine either due to the dispreference shown by PL towards such substrates, and moreover, because this dispreference was enhanced as a result of GSP.

The 5% GSP treatment was also tried on baked beef in further co-digestion experiments (Figure 3B). Interestingly, addition of GSP only resulted in decreased lipid digestibility by (12%; $p < 0.001$) when the digestion simulation was carried out with PL protocol. In the GL + PL digestion, no change was observed ($p = 0.316$). This shows that, in the case of a less marked presence of TAGs containing SCFA and MCFA, the decrease in gross lipolysis was less significant.

The effect of BTB was also tested with cream first. Diluted aqueous extract of a selected black tea cultivar with high tannin content was mixed with cream before digestion. In this case, cream to tea brew weight ratios of 1:1, 1:2 and 1:3 were tested (Figure 3C). Maximum lipolysis inhibitory effect of BTB was reached at the 1:2 level and a further addition did not increase the effect. The observed degree of lipolysis inhibition did not differ between the GL + PL and PL digestion simulations, 23% and 22% ($p = 0.067$) decrease in overall lipid digestibility was measured, respectively. Similarly to the experiments observed with GSP, no effect of BTB on overall lipid digestibility was observed with baked beef as a test food, when the same BTB to fat ratio was tested, which seemed effective for significant lipolysis inhibition in the case of the 1 cream: 2 tea experiment (Figure 3D). This result clearly suggests that differences in the fat source, i.e., the type and form of substrates for digestive lipase enzymes had a marked influence on the impact of the same fat-free test food on lipolysis. Most probably, the difference in FA profile, TAG structure and other matrix specific characteristics of the consumed food not only influences lipid digestibility of fat from different foods, but also influenced the lipolysis inhibitory effects of the fat-free test foods when consumed together with substrate test foods containing different types of fat sources. Lipase inhibitory properties of black tea are often attributed to theaflavins and it was also speculated that theaflavins bind to both the free enzyme as well as the enzyme-substrate complex, and likely at a site other than the substrate-binding site [48]. However, our above observation indicates that the lipolysis inhibitory property of black tea is more complex under physiologically more realistic conditions, and it can be sensitively modulated by the microstructure of the food [67].

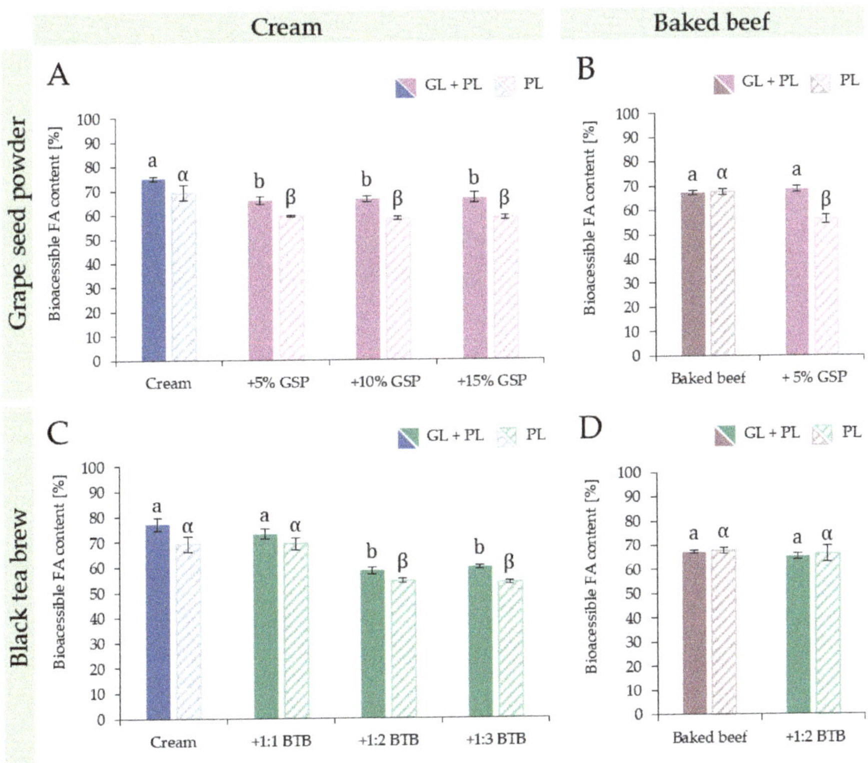

Figure 3. Dose dependency tests with cream test food (**A**,**C**) and effect of foods with bioactive constituents on baked beef (**B**,**D**) in the most effective ratio. (**A**,**B**): effect of grape seed powder (GSP); (**C**,**D**): effect of black tea brew (BTB). Significant difference within groups is marked with Latin letters (GL + PL) or with Greek letters (PL) ($p < 0.05$).

3.3.2. Fatty Acid Specific Effects: Reduced Digestibility of SCFAs and MCFAs

Our results showed that the exerted lipolysis inhibitory effects of the tested GSP and BTB were substrate-dependent, i.e., the given amount that was sufficient to restrict cream lipid digestion was not enough to provide the same impact for baked beef test food with identical lipid concentration. To further study the impact of fat-free test foods, FA-specific analysis was conducted (Figures 4 and 5).

In the case of GSP and BTB treatments, a similar trend regarding the changes in FA profile could be observed for cream as a substrate test food. Co-digestion with GSP and BTB primarily affected the MCFAs. In GL + PL digestion simulations with GSP (Figure 4A), the release of SCFAs and MCFAs were reduced more drastically than the release of LCFAs. Namely, the digestibility of SCFAs and MCFAs (C8:0 C15:0) reduced in the range of 43–22% and the release of LCFAs were reduced between 5 and 18%, compared to the control digestions. In the case of BTB (Figure 5A), the inhibitory effect in digestion simulation with both enzymes were more evened out, since the reduction in the release of all FAs was between 16 and 50%; however, the average reduction was 5% higher for SCFAs and MCFAs. Moreover, the impact of both foods seemed to be more pronounced on the pancreatic lipase activity. After PL only digestions (Figures 4B and 5B), the addition of both GSP and BTB resulted in a decreased release of SCFAs and MCFAs. With GSP, RR of C10:0–C15:0 decreased by 46–32% compared to average 7% decrease of longer FAs, and with BTB, RR of C6:0 and C10:0–C15:0 decreased by 52% and 51–25%, respectively, compared to the average 19% of longer FAs.

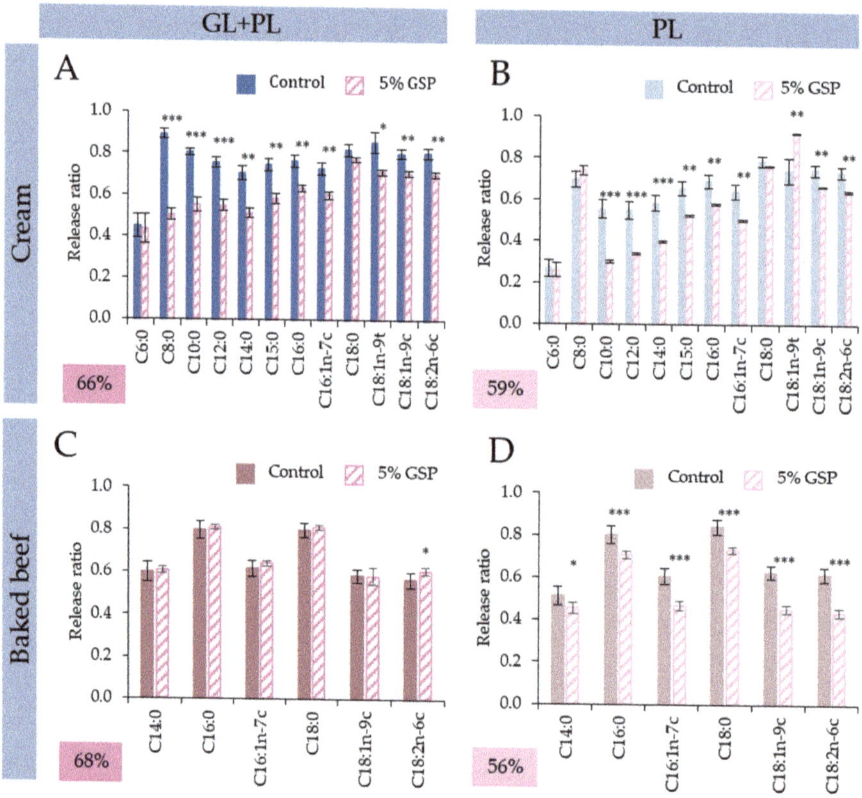

Figure 4. Effect of 5 w% of grape seed powder on release ratio of fatty acids. (**A**,**B**): cream test food in digestion simulation with GL + PL and PL, respectively; (**C**,**D**): baked beef test food in digestion simulation with GL + PL and PL, respectively. Asterisk labels (*) show significant difference between release ratio of individual fatty acids after GL + PL and PL digestions (* $p < 0.05$; ** $p < 0.01$; *** $p < 0.001$).

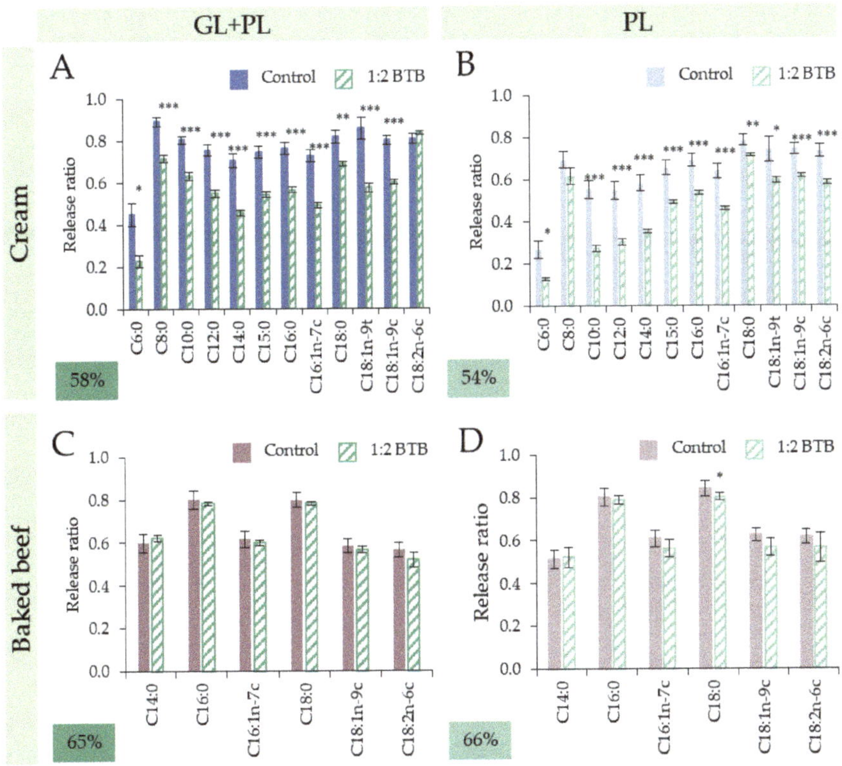

Figure 5. Effect of black tea brew (1:2 ratio) on release ratio of fatty acids. (**A**,**B**): cream test food in digestion simulation with GL + PL and PL, respectively; (**C**,**D**): baked beef test food in digestion simulation with GL + PL and PL, respectively. Asterisk labels (*) show significant difference between release ratio of individual fatty acids after GL + PL and PL digestions (* $p < 0.05$; ** $p < 0.01$; *** $p < 0.001$).

Interestingly one exception was found for BB, where lipolysis restriction was observed. Namely, when the 5% GSP was added and when only PL was used (Figure 3B). The observation could be attributed to the difference in the extent of the effect between the two fat-free test foods on PL, i.e., GSP had a higher effect on only PL than GL + PL and BTB did not. This effect was not uniform on all FAs (Figure 4D). The RR decrease was higher on the only MCFA (C14:0: 25%) and on the unsaturated FAs (C16:1n-7c: 27%, C18:1n-9c: 22% and C18:2n-6c: 29%) than on the other two, saturated FAs (C16:0: 13% and C18:0: 10%). This also highlights the specific non-preference of PL on the asymmetric TAGs, since the TAGs containing unsaturated FAs were also from this type of TAG (Figure 1).

4. Conclusions

In conclusion to the digestion simulation experiments carried out with two selected substrate test foods (cream and baked beef) having markedly different FA profiles, we postulate that TAGs containing SCFAs and MCFAs are non-preferred substrate types for pancreatic lipase (PL) and, consequently, PL alone is rather ineffective in the digestion of such TAGs containing SCFAs and MCFAs. However, this specificity was not characteristic for gastric lipase (GL). This effect was prominent in milk fat containing products such as the analyzed cream. The results also showed that when there was no marked diversity in FA composition of the consumed fat source and, consequently, heterogeneity in the TAG structure was limited, the contribution of GL could be negligible such as in the case of baked beef. We also found that the difference in FA profile, TAG structure and other matrix specific characteristics of the consumed food influenced the lipolysis inhibitory effects of the same fat-free food containing bioactive constituents, when consumed together with foods containing different types of fat sources.

Our findings together suggest that the lipolysis inhibitory effect of both GSP and BTB primarily affected the lipolysis of short and medium-chain FAs. The reason behind this can be explained by the hypothesis that the dispreference of PL to digest TAGs containing SCFAs and MCFAs was presumably further enhanced as a result of treatment with foods such as GSP or BTB. At the same time, we postulate that GL was also restricted to some extent and could not perform the pre-digestion of SCFA- and MCFA-containing TAGs efficiently either. It should be noted, however, that the reasons for the dispreference of PL towards TAGs containing SCFA and MCFA and the increase in dispreference when these lipids were consumed in a meal along with GSP or BTB was unclear.

Another interesting conclusion derived from this in vitro preliminary study is that the two different foods with bioactive constituents showed lipolysis-inhibiting effects with the same characteristics. That is, both GSP and BTB had a significant effect in the case of cream, while they proved to be ineffective in restricting the digestion of fat from baked beef. This finding highlights the importance of the characteristics of the fat source and the food matrix as key determinants in the observed lipolysis inhibitory effect of GSP and BTB. This observation also suggests that in future studies, the impact of various foods—rich in substances with proven lipase-inhibiting effect—on lipid digestion—should be ideally investigated with different types of real foods as substrates, which might specifically influence their presumed effects on lipolysis.

Author Contributions: Conceptualization, L.A. and J.T.; formal analysis, J.T.; data curation, J.T. and L.A.; writing—original draft preparation, J.T.; writing—review and editing, J.T. and L.A.; funding acquisition, L.A. All authors have read and agreed to the published version of the manuscript.

Funding: This research was funded by OTKA K135294, which is kindly acknowledged. JT acknowledges the financial support of the Doctoral School of Food Science at the Hungarian University of Agriculture and Life Sciences (MATE).

Institutional Review Board Statement: Not applicable.

Informed Consent Statement: Not applicable.

Data Availability Statement: Not applicable.

Acknowledgments: Tea sample was provided by Eszter Benes, which is kindly acknowledged.

Conflicts of Interest: The authors declare no conflict of interest.

References

1. Ng, M.; Fleming, T.; Robinson, M.; Thomson, B.; Graetz, N.; Margono, C.; Mullany, E.C.; Biryukov, S.; Abbafati, C.; Abera, S.F.; et al. Global, Regional, and National Prevalence of Overweight and Obesity in Children and Adults during 1980–2013: A Systematic Analysis for the Global Burden of Disease Study 2013. *Lancet* **2014**, *384*, 766–781. [CrossRef] [PubMed]
2. Tune, J.D.; Goodwill, A.G.; Sassoon, D.J.; Mather, K.J. Cardiovascular Consequences of Metabolic Syndrome. *Transl. Res.* **2017**, *183*, 57–70. [CrossRef] [PubMed]
3. Koliaki, C.; Liatis, S.; Kokkinos, A. Obesity and Cardiovascular Disease: Revisiting an Old Relationship. *Metabolism* **2019**, *92*, 98–107. [CrossRef]
4. World Health Organization; Regional Office for Europe. *WHO European Regional Obesity Report 2022*; World Health Organization: Geneva, Switzerland; Regional Office for Europe: Copenhagen, Denmark, 2022; ISBN 978-92-890-5773-8.
5. Cercato, C.; Fonseca, F.A. Cardiovascular Risk and Obesity. *Diabetol. Metab. Syndr.* **2019**, *11*, 74. [CrossRef]
6. Wang, L.; Wang, H.; Zhang, B.; Popkin, B.M.; Du, S. Elevated Fat Intake Increases Body Weight and the Risk of Overweight and Obesity among Chinese Adults: 1991–2015 Trends. *Nutrients* **2020**, *12*, 3272. [CrossRef] [PubMed]
7. Wu, C.; Mi, B.; Luo, W.; Chen, B.; Ma, J.; Huang, H.; Zhang, Q.; Wang, Y.; Liu, H.; Yan, B.; et al. Association between Long-Term Changes in Dietary Percentage of Energy from Fat and Obesity: Evidence from over 20 Years of Longitudinal Data. *Nutrients* **2022**, *14*, 3373. [CrossRef]

8. Forouhi, N.G.; Sharp, S.J.; Du, H.; van der A, D.L.; Halkjœr, J.; Schulze, M.B.; Tjønneland, A.; Overvad, K.; Jakobsen, M.U.; Boeing, H.; et al. Dietary Fat Intake and Subsequent Weight Change in Adults: Results from the European Prospective Investigation into Cancer and Nutrition Cohorts. *Am. J. Clin. Nutr.* **2009**, *90*, 1632–1641. [CrossRef]
9. Raatz, S.; Conrad, Z.; Johnson, L.; Picklo, M.; Jahns, L. Relationship of the Reported Intakes of Fat and Fatty Acids to Body Weight in US Adults. *Nutrients* **2017**, *9*, 438. [CrossRef]
10. Suara, S.B.; Siassi, F.; Saaka, M.; Foroshani, A.R.; Asadi, S.; Sotoudeh, G. Dietary Fat Quantity and Quality in Relation to General and Abdominal Obesity in Women: A Cross-Sectional Study from Ghana. *Lipids Health Dis.* **2020**, *19*, 67. [CrossRef]
11. Beulen, Y.; Martínez-González, M.; van de Rest, O.; Salas-Salvadó, J.; Sorlí, J.; Gómez-Gracia, E.; Fiol, M.; Estruch, R.; Santos-Lozano, J.; Schröder, H.; et al. Quality of Dietary Fat Intake and Body Weight and Obesity in a Mediterranean Population: Secondary Analyses within the PREDIMED Trial. *Nutrients* **2018**, *10*, 2011. [CrossRef]
12. Ockner, R.K.; Manning, J.A. Fatty acid-binding protein in small intestine identification, isolation, and evidence for its role in cellular fatty acid transport. *J. Clin. Investig.* **1974**, *54*, 326–338. [CrossRef] [PubMed]
13. Mansbach, C.M.; Gorelick, F. Development and Physiological Regulation of Intestinal Lipid Absorption. II. Dietary Lipid Absorption, Complex Lipid Synthesis, and the Intracellular Packaging and Secretion of Chylomicrons. *Am. J. Physiol. Gastrointest. Liver Physiol.* **2007**, *293*, G645–G650. [CrossRef] [PubMed]
14. Iqbal, J.; Hussain, M.M. Intestinal Lipid Absorption. *Am. J. Physiol. Endocrinol. Metab.* **2009**, *296*, E1183–E1194. [CrossRef] [PubMed]
15. Mu, H. The Digestion of Dietary Triacylglycerols. *Prog. Lipid Res.* **2004**, *43*, 105–133. [CrossRef]
16. Lichtenstein, A.H.; Jones, P.J.H. Lipids: Absorption and Transport. In *Present Knowledge in Nutrition*; Erdman, J.W., Macdonald, I.A., Zeisel, S.H., Eds.; Wiley: Hoboken, NJ, USA, 2012; pp. 118–131. ISBN 978-0-470-95917-6.
17. Armand, M. Lipases and Lipolysis in the Human Digestive Tract: Where Do We Stand? *Curr. Opin. Clin. Nutr. Metab. Care* **2007**, *10*, 156–164. [CrossRef]
18. Liu, T.-T.; Liu, X.-T.; Chen, Q.-X.; Shi, Y. Lipase Inhibitors for Obesity: A Review. *Biomed. Pharmacother.* **2020**, *128*, 110314. [CrossRef]
19. de la Garza, A.; Milagro, F.; Boque, N.; Campión, J.; Martínez, J. Natural Inhibitors of Pancreatic Lipase as New Players in Obesity Treatment. *Planta Med.* **2011**, *77*, 773–785. [CrossRef]
20. Kumar, A.; Chauhan, S. Pancreatic Lipase Inhibitors: The Road Voyaged and Successes. *Life Sci.* **2021**, *271*, 119115. [CrossRef]
21. Rajan, L.; Palaniswamy, D.; Mohankumar, S.K. Targeting Obesity with Plant-Derived Pancreatic Lipase Inhibitors: A Comprehensive Review. *Pharmacol. Res.* **2020**, *155*, 104681. [CrossRef]
22. Bustanji, Y.; Issa, A.; Mohammad, M.; Hudaib, M.; Tawah, K.; Alkhatib, H.; Almasri, I.; Al-khalidi, B. Inhibition of Hormone Sensitive Lipase and Pancreatic Lipase by Rosmarinus Officinalis Extract and Selected Phenolic Constituents. *J. Med. Plants Res.* **2010**, *4*, 2235–2242. [CrossRef]
23. Ninomiya, K.; Matsuda, H.; Shimoda, H.; Nishida, N.; Kasajima, N.; Yoshino, T.; Morikawa, T.; Yoshikawa, M. Carnosic Acid, a New Class of Lipid Absorption Inhibitor from Sage. *Bioorg. Med. Chem. Lett.* **2004**, *14*, 1943–1946. [CrossRef] [PubMed]
24. Yu, X.; Wang, X.-P.; Lei, F.; Jiang, J.-F.; Li, J.; Xing, D.-M.; Du, L.-J. Pomegranate Leaf Attenuates Lipid Absorption in the Small Intestine in Hyperlipidemic Mice by Inhibiting Lipase Activity. *Chin. J. Nat. Med.* **2017**, *15*, 732–739. [CrossRef] [PubMed]
25. Moreno, D.A.; Ilic, N.; Poulev, A.; Brasaemle, D.L.; Fried, S.K.; Raskin, I. Inhibitory Effects of Grape Seed Extract on Lipases. *Nutrition* **2003**, *19*, 876–879. [CrossRef]
26. Dwibedi, V.; Jain, S.; Singhal, D.; Mittal, A.; Rath, S.K.; Saxena, S. Inhibitory Activities of Grape Bioactive Compounds against Enzymes Linked with Human Diseases. *Appl. Microbiol. Biotechnol.* **2022**, *106*, 1399–1417. [CrossRef]
27. Nakai, M.; Fukui, Y.; Asami, S.; Toyoda-Ono, Y.; Iwashita, T.; Shibata, H.; Mitsunaga, T.; Hashimoto, F.; Kiso, Y. Inhibitory Effects of Oolong Tea Polyphenols on Pancreatic Lipase in Vitro. *J. Agric. Food Chem.* **2005**, *53*, 4593–4598. [CrossRef] [PubMed]
28. Chen, T.-Y.; Wang, M.M.C.; Hsieh, S.-K.; Hsieh, M.-H.; Chen, W.-Y.; Tzen, J.T.C. Pancreatic Lipase Inhibition of Strictinin Isolated from Pu'er Tea (Cammelia Sinensis) and Its Anti-Obesity Effects in C57BL6 Mice. *J. Funct. Foods* **2018**, *48*, 1–8. [CrossRef]
29. Maicas, S.; Mateo, J.J. Sustainability of Wine Production. *Sustainability* **2020**, *12*, 559. [CrossRef]
30. Ferrer-Gallego, R.; Silva, P. The Wine Industry By-Products: Applications for Food Industry and Health Benefits. *Antioxidants* **2022**, *11*, 2025. [CrossRef]
31. Akaberi, M.; Hosseinzadeh, H. Grapes (Vitis Vinifera) as a Potential Candidate for the Therapy of the Metabolic Syndrome. *Phytother. Res.* **2016**, *30*, 540–556. [CrossRef]
32. Serea, D.; Condurache, N.N.; Aprodu, I.; Constantin, O.E.; Bahrim, G.-E.; Stănciuc, N.; Stanciu, S.; Rapeanu, G. Thermal Stability and Inhibitory Action of Red Grape Skin Phytochemicals against Enzymes Associated with Metabolic Syndrome. *Antioxidants* **2022**, *11*, 118. [CrossRef]
33. Saucier, C.; Mirabel, M.; Daviaud, F.; Longieras, A.; Glories, Y. Rapid Fractionation of Grape Seed Proanthocyanidins. *J. Agric. Food Chem.* **2001**, *49*, 5732–5735. [CrossRef] [PubMed]
34. Das, A.K.; Islam, M.N.; Faruk, M.O.; Ashaduzzaman, M.; Dungani, R. Review on Tannins: Extraction Processes, Applications and Possibilities. *S. Afr. J. Bot.* **2020**, *135*, 58–70. [CrossRef]
35. Chedea, V.S.; Echim, C.; Braicu, C.; Andjelkovic, M.; Verhe, R.; Socaciu, C. Composition in Polyphenols and Stability of the Aqueous Grape Seed Extract from the Romanian Variety "Merlot Recas". *J. Food Biochem.* **2011**, *35*, 92–108. [CrossRef]

36. Ramos-Pineda, A.M.; Carpenter, G.H.; García-Estévez, I.; Escribano-Bailón, M.T. Influence of Chemical Species on Polyphenol–Protein Interactions Related to Wine Astringency. *J. Agric. Food Chem.* **2020**, *68*, 2948–2954. [CrossRef]
37. Lucarini, M.; Durazzo, A.; Kiefer, J.; Santini, A.; Lombardi-Boccia, G.; Souto, E.; Romani, A.; Lampe, A.; Ferrari Nicoli, S.; Gabrielli, P.; et al. Grape Seeds: Chromatographic Profile of Fatty Acids and Phenolic Compounds and Qualitative Analysis by FTIR-ATR Spectroscopy. *Foods* **2019**, *9*, 10. [CrossRef]
38. Chantre, P.; Lairon, D. Recent Findings of Green Tea Extract AR25 (Exolise) and Its Activity for the Treatment of Obesity. *Phytomedicine* **2002**, *9*, 3–8. [CrossRef]
39. Gondoin, A.; Grussu, D.; Stewart, D.; McDougall, G.J. White and Green Tea Polyphenols Inhibit Pancreatic Lipase in Vitro. *Food Res. Int.* **2010**, *43*, 1537–1544. [CrossRef]
40. Glisan, S.L.; Grove, K.A.; Yennawar, N.H.; Lambert, J.D. Inhibition of Pancreatic Lipase by Black Tea Theaflavins: Comparative Enzymology and In Silico Modeling Studies. *Food Chem.* **2017**, *216*, 296–300. [CrossRef]
41. Sellami, M.; Louati, H.; Kamoun, J.; Kchaou, A.; Damak, M.; Gargouri, Y. Inhibition of Pancreatic Lipase and Amylase by Extracts of Different Spices and Plants. *Int. J. Food Sci. Nutr.* **2017**, *68*, 313–320. [CrossRef]
42. Zhang, H.; Qi, R.; Mine, Y. The Impact of Oolong and Black Tea Polyphenols on Human Health. *Food Biosci.* **2019**, *29*, 55–61. [CrossRef]
43. Zhang, R.; Xing, D.; Wang, C. Pancreatic Triglyceride Lipase Inhibitors Derived from Natural Products: How To Dig into the Truth. *J. Agric. Food Chem.* **2021**, *69*, 6097–6099. [CrossRef] [PubMed]
44. Tan, Y.; Zhang, Z.; Zhou, H.; Xiao, H.; McClements, D.J. Factors Impacting Lipid Digestion and β-Carotene Bioaccessibility Assessed by Standardized Gastrointestinal Model (INFOGEST): Oil Droplet Concentration. *Food Funct.* **2020**, *11*, 7126–7137. [CrossRef] [PubMed]
45. Tan, Y.; Zhang, Z.; Liu, J.; Xiao, H.; McClements, D.J. Factors Impacting Lipid Digestion and Nutraceutical Bioaccessibility Assessed by Standardized Gastrointestinal Model (INFOGEST): Oil. *Food Funct.* **2020**, *11*, 9936–9946. [CrossRef] [PubMed]
46. Tan, Y.; Zhou, H.; McClements, D.J. Application of Static in Vitro Digestion Models for Assessing the Bioaccessibility of Hydrophobic Bioactives: A Review. *Trends Food Sci. Technol.* **2022**, *122*, 314–327. [CrossRef]
47. Verkempinck, S.H.E.; Salvia-Trujillo, L.; Moens, L.G.; Carrillo, C.; Van Loey, A.M.; Hendrickx, M.E.; Grauwet, T. Kinetic Approach to Study the Relation between in Vitro Lipid Digestion and Carotenoid Bioaccessibility in Emulsions with Different Oil Unsaturation Degree. *J. Funct. Foods* **2018**, *41*, 135–147. [CrossRef]
48. Majeed, H.; Antoniou, J.; Hategekimana, J.; Sharif, H.R.; Haider, J.; Liu, F.; Ali, B.; Rong, L.; Ma, J.; Zhong, F. Influence of Carrier Oil Type, Particle Size on in Vitro Lipid Digestion and Eugenol Release in Emulsion and Nanoemulsions. *Food Hydrocoll.* **2016**, *52*, 415–422. [CrossRef]
49. Calvo-Lerma, J.; Asensio-Grau, A.; Heredia, A.; Andrés, A. Screening the Impact of Food Co-Digestion on Lipolysis under Sub-Optimal Intestinal Conditions. *LWT* **2020**, *118*, 108792. [CrossRef]
50. Brodkorb, A.; Egger, L.; Alminger, M.; Alvito, P.; Assunção, R.; Ballance, S.; Bohn, T.; Bourlieu-Lacanal, C.; Boutrou, R.; Carrière, F.; et al. INFOGEST Static In Vitro Simulation of Gastrointestinal Food Digestion. *Nat. Protoc.* **2019**, *14*, 991–1014. [CrossRef]
51. *MSZ 20685:1980*; Tea Fizikai És Kémiai Vizsgálata (Physico-Chemical Evaluation of Tea). Magyar Szabványügyi Hivatal: Budapest, Hungary, 1980.
52. *ISO 1444:2000*; Meat and Meat Products. Determination of Free Fat Content. ISO: Geneva, Switzerland, 2000.
53. *ISO 2450:2008*; Cream—Determination of Fat Content—Gravimetric Method (Reference Method). ISO: Geneva, Switzerland, 2009.
54. Tormási, J.; Abrankó, L. Assessment of Fatty Acid-Specific Lipolysis by In Vitro Digestion and GC-FID. *Nutrients* **2021**, *13*, 3889. [CrossRef]
55. Bligh, E.G.; Dyer, W.J. A rapid method of total lipid extraction and purification. *Can. J. Biochem. Physiol.* **1959**, *37*, 911–917. [CrossRef]
56. *EN ISO 12966-2:2017*; Animal and Vegetable Fats and Oils—Gas Chromatography of Fatty Acid Methyl Esters—Part 2: Preparation of Methyl Esters of Fatty Acids. ISO: Geneva, Switzerland, 2017.
57. Omar, K.A.; Gounga, M.E.; Liu, R.; Mwinyi, W.; Aboshora, W.; Ramadhan, A.H.; Sheha, K.A.; Wang, X. Triacylglycerol Composition, Melting and Crystallization Profiles of Lipase Catalysed Anhydrous Milk Fats Hydrolysed. *Int. J. Food Prop.* **2017**, *20*, 1230–1245. [CrossRef]
58. Angers, P.; Tousignant, É.; Boudreau, A.; Arul, J. Regiospecific Analysis of Fractions of Bovine Milk Fat Triacylglycerols with the Same Partition Number. *Lipids* **1998**, *33*, 1195–1201. [CrossRef] [PubMed]
59. Christie, W.W.; Clapperton, J.L. Structures of the Triglycerides of Cows' Milk, Fortified Milks (Including. Infant Formulae), and Human Milk. *Int. J. Dairy Technol.* **1982**, *35*, 22–24. [CrossRef]
60. Smith, S.B.; Yang, A.; Larsen, T.W.; Tume, R.K. Positional Analysis of Triacylglycerols from Bovine Adipose Tissue Lipids Varying in Degree of Unsaturation. *Lipids* **1998**, *33*, 197–207. [CrossRef]
61. Infantes-Garcia, M.R.; Verkempinck, S.H.E.; Guevara-Zambrano, J.M.; Andreoletti, C.; Hendrickx, M.E.; Grauwet, T. Enzymatic and Chemical Conversions Taking Place during in Vitro Gastric Lipid Digestion: The Effect of Emulsion Droplet Size Behavior. *Food Chem.* **2020**, *326*, 126895. [CrossRef] [PubMed]
62. Carrière, F.; Rogalska, E.; Cudrey, C.; Ferrato, F.; Laugier, R.; Verger, R. In Vivo and in Vitro Studies on the Stereoselective Hydrolysis of Tri- and Diglycerides by Gastric and Pancreatic Lipases. *Bioorg. Med. Chem.* **1997**, *5*, 429–435. [CrossRef]

63. Mitchell, D.; Rodriguez, J.; Carriere, F.; Krieger, N. Determination of the Quantitative Stereoselectivity Fingerprint of Lipases during Hydrolysis of a Prochiral Triacylglycerol. *J. Biotechnol.* **2008**, *135*, 168–173. [CrossRef]
64. Benito-Gallo, P.; Franceschetto, A.; Wong, J.C.M.; Marlow, M.; Zann, V.; Scholes, P.; Gershkovich, P. Chain Length Affects Pancreatic Lipase Activity and the Extent and PH–Time Profile of Triglyceride Lipolysis. *Eur. J. Pharm. Biopharm.* **2015**, *93*, 353–362. [CrossRef]
65. Hur, S.J.; Lim, B.O.; Park, G.B.; Joo, S.T. Effects of Various Fiber Additions on Lipid Digestion during In Vitro Digestion of Beef Patties. *J. Food Sci.* **2009**, *74*, C653–C657. [CrossRef]
66. Reis, P.; Holmberg, K.; Watzke, H.; Leser, M.E.; Miller, R. Lipases at Interfaces: A Review. *Adv. Colloid Interface Sci.* **2009**, *147–148*, 237–250. [CrossRef]
67. Lamothe, S.; Guérette, C.; Britten, M. Nutrient Release and Oxidative Stability during in Vitro Digestion of Linseed Oil Emulsions Produced from Cow Milk, Soy Drink, and Green Tea Extract. *LWT* **2020**, *134*, 110137. [CrossRef]

MDPI AG
Grosspeteranlage 5
4052 Basel
Switzerland
Tel.: +41 61 683 77 34

Nutrients Editorial Office
E-mail: nutrients@mdpi.com
www.mdpi.com/journal/nutrients

www.ingramcontent.com/pod-product-compliance
Lightning Source LLC
Chambersburg PA
CBHW062056220926
51850CB00010B/2789
* 9 7 8 3 7 2 5 8 6 0 3 5 7 *